rüffer & rub

Essentials of Deductive Physics

Foundation for the Great Theories of Physics

Hans Widmer

English translation by
Richard Rabenkamp

First edition 2014

info@ruefferundrub.ch | www.ruefferundrub.ch

Photograph of Hans Widmer, p. 138: Felix Ghezzi

Printed: Books on Demand GmbH, Norderstedt
Paper: Cream white, 90 g/m²

ISBN: 978-3-907625-70-5

Content

Preface to the English Edition

Deductive Physics does arguably lead to the established laws of physics; however, it is totally heterodox, pole-axes the traditional physicist and might easily be dismissed as yet another "fantastic idea". That is why a short introduction titled "What is Gained with 'Deductive Physics'?" precedes the translation from German. In addition, the reader's attention may be drawn to the fact that deductive physics is the continuation of the works of Lorentz[1853–1928] and Poincaré[1854–1912], which Einstein explicitly has paid tribute to[1955].

On the occasion of a lecture at the Faculty of the Department of Physics at ETH Zurich abbreviated derivations of 15 laws for which conventional inductive physics cannot provide reason—save that they stand the test of time—were presented.

What is Gained with "Deductive Physics"?

Towards the end of the seminar at ETH Zurich on 15 May 2014 Gianni Blatter, professor of theoretical physics, asked the crucial questions: What is gained with this physics, which open items does it resolve or what does it predict?

Well, in general physics would be simplified by one order of magnitude:

- the closed theories of *inductive physics* comprise 75 independent laws and constants, to these come the uncounted ones of super symmetry and string theory,
- while deductive physics derives all laws from its single axiom, the specific continuum (with c, G, $\hbar$); with it all explanations are monolithic and there is no need for further grand unification.

I – On the Matter of Physics

Relativity Theory (RT)

1. Inertia does not require any Higgs boson for the field needed to represent it: the influx towards mass *is* this field. The phenomenon "mass" results from interactions with fields outside the hadron. It determines all frequencies within the hadron, though there is no mass per se inside the hadron.
2. Gravitation does not require any graviton (the latter merely being postulated in analogy to the virtual photon), but results from mutual attraction of sources and sinks, which the phenomenon "mass" is both at the same time (the constituents of the continuum could be called gravitons).
3. Dark matter is not needed and the miracle of what it is vanishes. The attraction towards the centres of galaxies is explained by their black hole—however, not by attractive force, but by floating along the influx towards the black hole.
4. Dark energy is not needed and the miracle of what it is vanishes. The constantly growing black holes, which repel one another as sinks, explain the accelerated expansion of the universe. Friedmann models are impossible (a primordial concentration with infinite mass density and vanishing size are ruled out by the constituents of the continuum occupying space; singularities altogether are impossible); the cosmological constant is useless.
5. The universe has more degrees of freedom than represented in the standard model: G and $\hbar$ change as the density of the conti-

nuum decreases. Only *c* remains constant (as the relation of space and time in subjective perception).

Quantum Mechanics (QM)

1. The probability laws of QM are consequences of continuum dynamics. They are not the fundamental laws of nature, forcing departure from causality.
2. The new conservation quantity QM introduces is not only hidden in the Schrödinger equation and is not a mere quantum number, but is the field energy annihilated by the interference wave.
3. The wave QM introduces is not primarily the probability amplitude for the locus of a particle (which often leads to false conclusions: for instance that the e^- would mainly be occupying the centre of the H_1^1), but is the interference wave of the radiations, exuding from masses.
4. The action quantum is not a quantum of anything, but a property of the continuum (correlated with its free path length), and manifests itself in resonances. Wherever $\hbar$ appears, there are resonances.
5. Tuned states are based on synchronisation of mechanical frequencies and frequencies of interference waves from Lorentz contractions (for instance of the e^- inside H_1^1).
6. The electron inside the hydrogen atom oscillates, but does not radiate, because the corresponding radiation is absorbed within the interference wave.
7. The zero point energy does not come from nowhere, but is the annihilated radiation having turned into oscillation (residence at the origin requiring energy for which there is no source).
8. The Schrödinger equation arises from the frequency of superimposed Lorentz contractions from relative velocity and potential field at once.
9. Quantum entanglement (Einstein: "spooky action at a distance") rests on states tuning first, which the local and realistic field of the continuum can describe, whereas the tuned interference

wave *simultaneously* appears and can be measured in the entire space.

Standard Model of Elementary Particle Physics (SM)

1. All "quarks" are variations of one and the same vortex with different angular momenta, in different distances and different topologies.
2. Primary conservation quantity is the angular momentum (resonance in vortices → quark). When several vortices aggregate, there are two relative positions relative to each other: d and u. Topologically only c and t can take the place of u, that of d only s and b.
3. In the baryon three individually unviable vortices aggregate orthogonally (which neutralises attraction or repulsion from rotation). The three colours of the standard model are the three spatial axes, also appearing in the Dirac matrix. Leptons are composed tri-orthogonally as well.
4. Secondary conservation quantity is the topology of the baryon: the beta decay for instance is based on the position exchange of a vortex. All conservation quantities like parity refer to combinations of conservation of angular momentum and topology of the affected vortices. The gauge theories central in the SM define topological invariants.
5. The electroweak force is not a force, but reflects the small probability for topological conversion, which in turn rests on the large conversion energy required (6 new vortices provided by the W^- boson).
6. The strong interaction is based on phase shift by π of the radiation emitted from the vortices. *Confinement* represents an energy minimum by the interference wave. Higher energy particles consist of vortices in minima $n > 1$.
7. The asymmetry in the abundance of particles and anti-particles is caused by vortices being formed first in the Big Bang (total an-

gular momentum equals zero), then aggregating into particles, not anti-particles.

Electromagnetism

1. The electric field is generated by elementary particle dynamics ("charge" is a dynamics—nothing material). The identifying quantity is the fine structure constant.
2. Radiation has the rest frequency of the electron and is being emitted from a sphere having a radius of the Compton length.
3. Virtual photons are rotating waves with the Compton length as their radius (from resonance).
4. Electric forces are based on annihilation or repulsion of rotational radiation.
5. The Maxwell equations are the continuity equation for the vortex density B (magnetic field), the Euler equation for the vortex flux E (electric field).
6. The gauge theories employed in the SM are generalizations of the Maxwell equations, which essentially process rotation mathematically. Thus gauge theory turns out as the vehicle with which the SM introduces rotation.
7. Spin $\hbar/2$ of the Dirac particle results from the precession that the rotation of its three vortices causes mutually.

II – Epistemologically

1. Reduction of the number of independent constants and laws by about one order of magnitude.
2. Saving the effort of *physics by extrapolation*.
3. Being conscious that physics resides in the human brain:
   - space and time remain the coordinate system within which reason maps its experiences,
   - the continuum is the substrate in the human coordinate system with which physical phenomena can be emulated,

- laws are to be invented freely (Einstein)—not to be found in nature like hidden Easter eggs,
- the reconstruction of causality remains the objective: deductive physics studies "realistic and local" causes—while QM studies the statistics of outcomes,
- uncertainty reflects the lack of knowledge.

III – Didactically

1. Immovable foundation – physics in its most delicate ramifications and sublime mathematical complexities never contains more than what is embedded in its foundation; never has a great contribution to physics originated from intricate thinking.
2. *Grasping* the fundamentals of all physics in, say, twelve lectures and twelve tutorials by students with basic knowledge of calculus.

Principle of Deductive Physics

The laws of physics resulted from analysis of individual aspects of the material world; experiment and observation have always been the basis, which is why no common foundation arose for relativity, quantum mechanics, elementary particle physics and electromagnetism. As new phenomena arose which current understanding did not explain, one would turn to new concepts and increase the number of independent laws and constants: the completed theories comprise 75 of these, the still open ones over a hundred.[1] This dissatisfying state generates a permanent endeavour for unification ("Grand Unification", "Theory of Everything" etc.).

The great theories of physics are field theories, and this means: infinitesimal contact between causes and effects across all space; this was already the function of Descartes' ether. Hidden in all field theories there is a continuum, it is the assumption whose mathematics are field equations. While conventional physics proceeds inductively and concludes a continuum from phenomena, deductive physics (epistemological foundations[2]) pursues the reverse path, deriving from a continuum all material phenomena as dynamics of it:

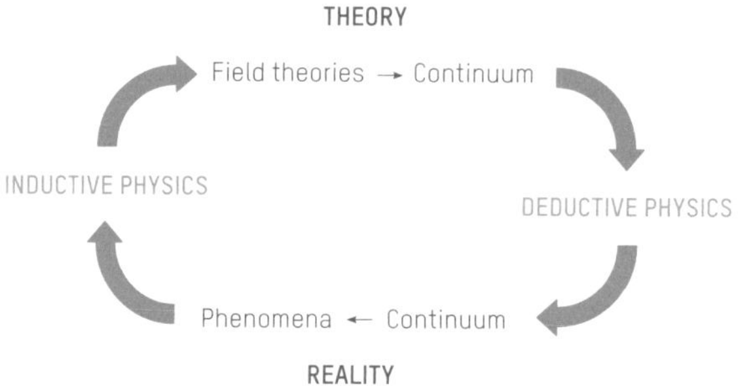

Continuum is the substrate organising itself into matter. Matter is *organisation* of continuum, not continuum. Elementary material organisations are the substrate for higher organisations. Inertia, gravitation, electromagnetism develop dynamically, they are not already contained in the elementary (emergence). The infinitesimal dynamics of the continuum is the foundation of deductive physics.

The constituents of this continuum are pure bodies, defined as permanent, impermeable volumes—counterparts to empty space and not distinguished by anything else (especially *not* by inertia: this results only from the *dynamics* of the continuum). Deductive physics requires only this for representing the material world: the coordinate system spanned by space and time as well as bodies therein as constituents of a specific continuum. This continuum is not just an idea as in Descartes nor does it hide in mathematics as in Einstein, but it:

- *is* the analogue to an isothermal gas, as already implied by the equations of the theory of relativity (TR);
- *is* specified by fundamental constants:
  - c for perturbation propagation speed, $c = \sqrt{\partial p/\partial \rho}$ ("speed of sound" within continuum),
  - G for inverse density (indirectly),
  - $\hbar$ for free path length (indirectly);

- *explains* what inductive physics simply accepts, specifically why
  - interactions are not instantaneous,
  - gravitation and electromagnetism have the same speed of propagation,
  - interference at atomic scale produces quantum mechanical phenomena (frequency shift),
  - interference at Compton length produces the strong force (phase shift);
- *links* TR and quantum mechanics (QM), as the de Broglie-Einstein Relations of QM derive from TR's Lorentz contraction;
- *reconciles* QM with Einstein's postulate of a "local and realistic" theory (specifically in the interpretation of quantum entanglement[3]);
- *supports* both:
  - the dynamics of the elementary building blocks of matter,
  - the expansion of the universe.

It is not nature imposing itself with concepts and laws, instead these have to be invented as hypotheses and validated in experiment.[4] The touchstone for perception in deductive physics are the laws of inductive physics. Thus it is standing on the shoulders of giants from Kepler to the inventors of the Standard Model.

I – Theory of Relativity

1. Mass Dynamics

Mass dynamics explains the macroscopic phenomena inertia, gravitation, relativity—remote from elementary particle dynamics. Continuum flows towards a point in space and from there it is radiated, superimposed over the influx:

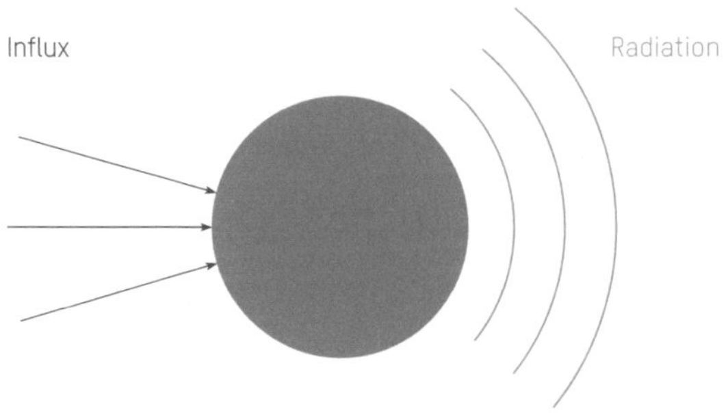

Inertia arises from the compression of the influx field (Lorentz contraction) with relative velocity of a mass against the resting continuum:

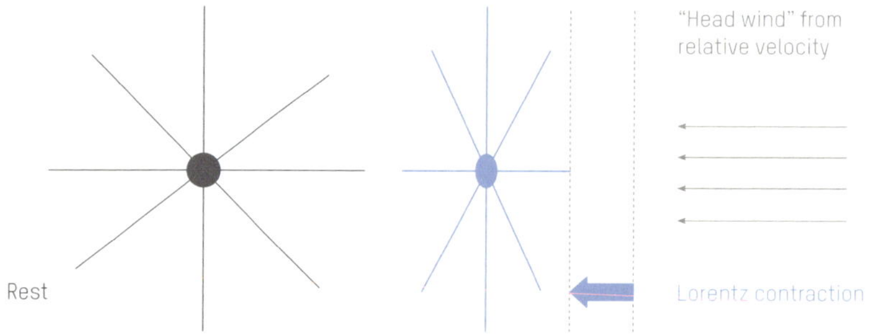

Gravitation arises from the impact of the outbound waves on the influx of distant masses, similar to attraction of sources and sinks in any continuum:

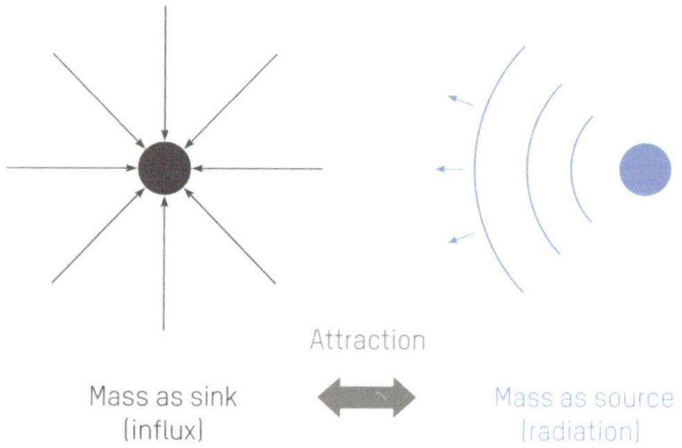

Two masses attract each other twice as source and sink, repel each other as two sinks once, do not interact as two sources. The result is one-fold attraction.

Equivalence principle: A mass cannot determine whether it has relative velocity against the continuum at rest or resides within the influx of another mass—the "head wind" is identical. Instead of kinetic energy, potential energy has to be entered into the formula for the Lorentz contraction, for instance $\sqrt{1-v^2/c^2} \rightarrow \sqrt{1-2GM/rc^2}$

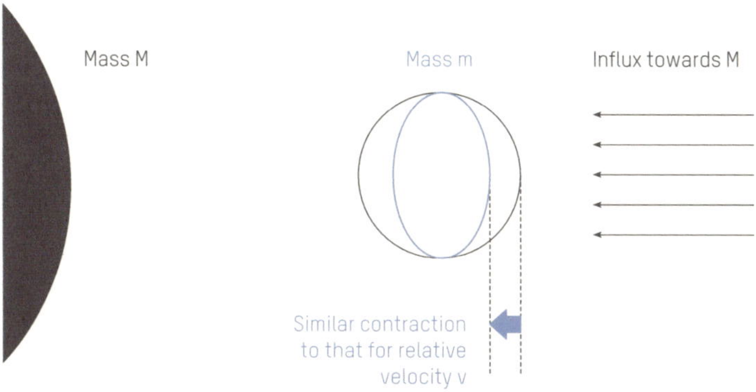

Theory of relativity: Its results are side effects of mass dynamics, specifically contraction of influx and radiation fields, leading to a shift of wavelengths and frequencies.

Cosmology: The drive for the expansion of the universe arises from the potential c^2 of the continuum.

Black holes form, if the outbound waves of a mass conglomerate can no longer overcome the influx; they permanently absorb continuum, the stars contained therein "float along".

Dark matter: The effect inductive physics ascribes to dark matter is generated by the influx towards the black hole within each galaxy.

Dark energy: The effect inductive physics ascribes to dark energy is generated by the black holes within each of the galaxies in our universe. These are sinks, and sinks repel each other.

The field equation is the basis for proving these claims. It is developed statistically for an infinitesimal space element (the momentum theorem is not available at this level of inspection) and results from linking the two axiomatically assumed, thus not provable, notions:

- Conservation of the quantity of continuum,
- Conservation of continuum movement (potential c^2).

2. Conservation of Quantity

In the interest of clarity the following considerations are made in just one dimension, the x-axis. The flow velocity along the x-axis is v. Density is denoted by ρ, although this symbol is usually applied to densities with unit kg/m^3, whereas density here is merely a coefficient, i.e. the portion of space occupied by the constituents.

At rest, flow across a boundary plane is $\rho c/2$ in one direction and the negative amount in the opposite. Here c represents the speed of light, the equivalent to speed of sound for real gas $c = \sqrt{\partial p / \partial \rho}$.

For a flow with velocity v the sum of the inbound and outbound fluxes into and out of the interval dx corresponds to

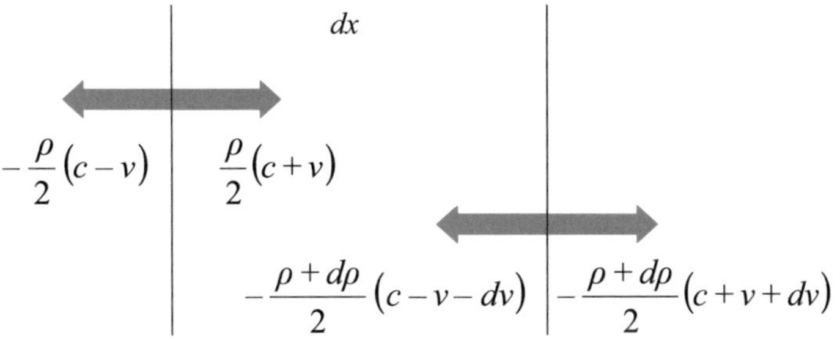

the change of density inside the interval

$$\rho dv + v d\rho + \frac{\partial \rho}{\partial t} dx = 0$$

From this follows the continuity equation C (the potential does not interfere)

$$C \quad \frac{\partial v}{\partial x} + \frac{v}{\rho}\frac{\partial \rho}{\partial x} + \frac{1}{\rho}\frac{\partial \rho}{\partial t} = 0$$

3. Conservation of Motion

In the following consideration the flow across the interval boundaries right and left remains equal and constant. If the density is reduced by $d\rho$ instantaneously at $x + dx$ and time $t = 0$, the continuum inside the interval acquires an average velocity component $dv/2$. Flux does remain constant, but is composed differently after the new equilibrium has been reached:

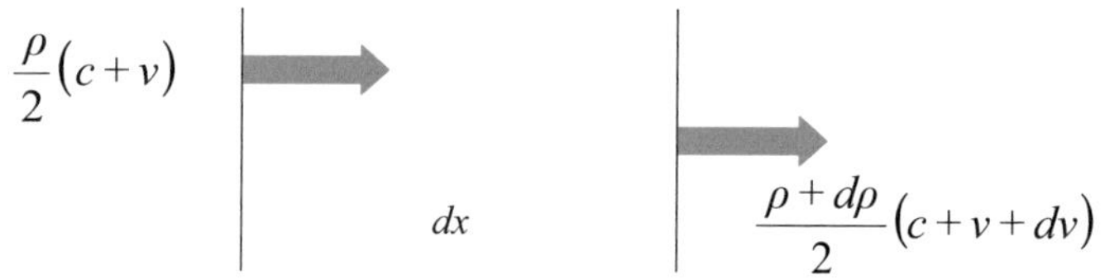

From equating both fluxes follows: $\rho dv + d\rho v + d\rho c = 0$. The new equilibrium is reached when the density reduction on the right has been conveyed to the left boundary, thus after $dt = dx/c$, from which the following equation results:

$$\rho\frac{dv}{dt} + c^2\frac{d\rho}{dx} + cv\frac{d\rho}{dx} = 0$$

The assumed velocity c of the "conveying" results from the wave equation.

With

- $v << c$ (neglecting the term on the right),
- $dv/dt = \partial v/\partial t + v\partial v/\partial x$,
- $d\rho/dx = \partial\rho/\partial x$ because of $\partial\rho/\partial t = 0$

the Euler equation M arises (here derived statistically, while conventionally the momentum theorem has to be applied twice):

$$M \quad \frac{\partial v}{\partial t} + v\frac{\partial v}{\partial x} + \frac{c^2}{\rho}\frac{\partial\rho}{\partial x} = 0$$

4. Field Equation

C and *M* form a system of equations for density and flow velocity (analogous to Maxwell equations). Instead of velocity v the potential $\Phi = \int v dx$ common in gas dynamics is used (x, path), as well as a new quantity instead of density ρ : $I = \ln \rho$: to simplify the mathematics. They now read as

$$
\begin{array}{ll}
C & \Phi_{xx} + \Phi_x I_x + I_t = 0 \\
M & \Phi_{xt} + \Phi_x \Phi_{xx} + c^2 I_x = 0
\end{array}
$$

Mutually inserting after differentiations and eliminating *I* yields the wanted field equation

- in one dimension

$$\Phi_{xx}\left(1 - \frac{\Phi_x^2}{c^2}\right) - \frac{2\Phi_x \Phi_{xt}}{c^2} - \frac{\Phi_{tt}}{c^2} = 0$$

- spherical (*r* radius)

$$\Phi_{rr}\left(1 - \frac{\Phi_r^2}{c^2}\right) + 2\Phi_r\left(\frac{1}{r} - \frac{\Phi_{rt}}{c^2}\right) - \frac{\Phi_{tt}}{c^2} = 0$$

The wave equation for continuum at rest Φ_r, $\Phi_x \ll$ c leaps directly from the field equation:

$$\Phi_{rr} - \frac{\Phi_{tt}}{c^2} = 0$$

This result legitimates the *c* used for deriving *M*. Analogously ensue propagation of electric field *E* and magnetic field *B*:

$$
\begin{array}{ll}
C & rot E + B_t = 0 \\
M & E_t - c^2 rot\, B = 0
\end{array}
\quad \rightarrow \quad \nabla^2 E - \frac{E_{tt}}{c^2} = 0
$$

With a flux present ($\Phi_x > 0$) x has to be replaced by $x' = \dfrac{x - \Phi_x t}{\sqrt{1 - \Phi_x^2 / c^2}}$

and the wave equation takes the simple form above.

The factor $1 - v^2/c^2$ (with $\Phi_x \rightarrow v$) representing the field contraction forms the basis for all relativistic phenomena.

Solution of the field equation of a spherically symmetrical dynamics: for large radii Φ is treated as superposition of a flow and a perturbation: $\Phi = \Phi^{flow} + \Phi^{perturbation}$. For large radii/small velocities follow

$$\nabla^2 \Phi^{flow} = 0$$

$$\nabla^2 \Phi^{perturbation} - \frac{\Phi_{tt}^{perturbation}}{c^2} = 0$$

The first represents an incompressible influx towards the centre, the second a radiation superimposed onto the influx with the solutions

$$\Phi^{flow} = \frac{\Phi_o^{flow}}{r}$$

$$\Phi^{perturbation} = \frac{\Phi_o^{perturbation} e^{-i(kr - \omega t)}}{r}.$$

These approximations suffice for determining inertia and gravitation. However, the “radius” of mass cannot be obtained from them—it is defined quantum mechanically and by 38 orders of magnitude larger than the result of the computation of r for $\Phi_r = c$. Influx velocity and amount scale as $1/r^2$, radiation amplitude as $1/r$ and radiation amount therefore as $1/r^2$, thus ensuring continuity at any radius.

5. Influx Computed Exactly

Since the influx is stationary, time derivatives drop out of the field equation; spherical field equation:

$$\Phi_{rr}\left(1-\frac{\Phi_r^2}{c^2}\right)+\Phi_r\left(\frac{2}{r}\right)=0$$

The closed solution for v (= Φ_r) contains *productLog*. The relation between $\Phi_r = v$ and r resulting from integration over r is concrete:

$$vr^2 = cr_o^2 e^{-\frac{1}{2}\left(1-\frac{v^2}{c^2}\right)}$$

If both sides are multiplied by density and subjected to the continuity rule, the density profile ensues (with $\rho_{r\rightarrow\infty} = \rho_\infty$):

$$\rho(r) = \rho_\infty e^{-v(r)^2/2c^2}$$

The radius at which influx is reflected as radiation is computed from the incompressible kinetic potential (linearised as in TR): if the potentials are equal at any radius,

$$\frac{Gm}{r}=\frac{v^2}{2}$$

then with the influx reaching c, $r = 2Gm/c^2$. This determines the radius of a black hole (Schwarzschild radius).

The compressibly computed inversion radius is $r_{inversion} = r_{Schwarzschild} \cdot e^{1/4}$; there $I = \ln \rho = 1/2 \rightarrow \rho(r_{Schwarzschild}) = e^{-1/2}\rho_\infty$ holds.

Only about $r_{inversion}$ does the compressible solution depart from the incompressible one.

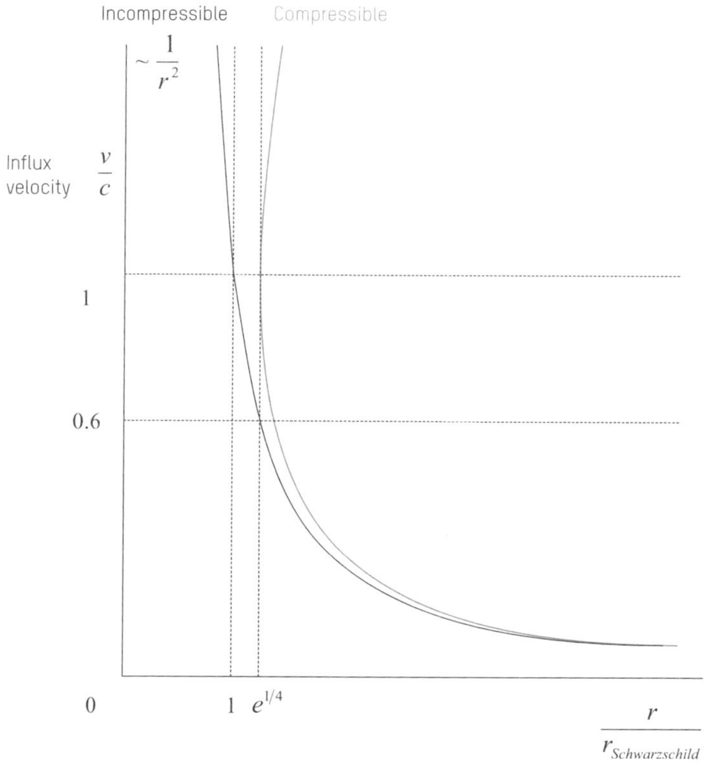
Incompressible
Compressible
$\sim \frac{1}{r^2}$
Influx velocity
$\frac{v}{c}$
1
0.6
0
1
$e^{1/4}$
$\frac{r}{r_{Schwarzschild}}$

6. Inertia

Inertia is based on

- the influx field of a mass within a counter flow being Lorentz contracted,
- this contraction offering resistance.

Thus Lorentz contraction rises above the shadow of theory of relativity to the quintessential element for explaining inertia. For inertia influx alone is relevant: radiation passes over all flowing with relative velocity c.

The field equation for the influx field of a sink within a counter flow has to be written in cylindrical coordinates for the one dimensional axial counter flow to enter into the calculation in a simple way. It is to be looked up in a text book on gas dynamics. For isothermal, vortex free situations it reads

$$\Phi_{xx}\left(1-\frac{\Phi_x^2}{c^2}\right)+\Phi_{rr}\left(1-\frac{\Phi_r^2}{c^2}\right)-\frac{2\Phi_x\,\Phi_r}{c^2}\,\Phi_{xr}+\frac{\Phi_r}{r}=0$$

(Counter flow on x-axis, r orthogonal to it)

With $\Phi_r \ll c$ this field equation reduces to

$$\Phi_{xx}\left(1-\frac{\Phi_x^2}{c^2}\right)+\frac{1}{r}\left(\Phi_r\, r\right)_r=0.$$

Here

- $\frac{\Phi_x^2}{c^2}$ represents compression,
- $\frac{1}{r}\left(\Phi_r r\right)_r$ represents the divergence of the incompressible case.

In the next step the counter flow has to be superimposed onto the flow field of the sink. The resulting total potential Φ^* consists of the one of the counter flow vx and the one of the sink $\Phi : \Phi^* = vx + \Phi$. Inserted into the field equation:

$$\Phi_{xx}\left(1-\frac{1}{c^2}\left(v+\Phi_r\right)^2\right)+\frac{\Phi_r}{r}+\Phi_{rr}\left(1-\frac{\Phi_r^2}{c^2}\right)$$

$$-\frac{2\Phi_r}{c^2}\left(v+\Phi_r\right)\Phi_{rx}=0$$

For $\Phi_r << v$ (lean bodies) the following field equation for the influx field ensues:

$$\Phi_{xx}\left(1-\frac{v^2}{c^2}\right)+\frac{1}{r}\left(\Phi_r r\right)_r=0$$

With $x' = \frac{x-vt}{\sqrt{1-v^2/c^2}}$ the equation becomes

$$\Phi_{x'x'}+\frac{1}{r}\left(\Phi_r r\right)_r=0, \text{ with solution}$$

$$\Phi=\frac{\Phi_o}{\sqrt{x'^2+r^2}}, \text{ written out } \Phi=\frac{\Phi_o}{\sqrt{\frac{(x-vt)^2}{1-v^2/c^2}+r^2}}$$

For a companion on the x-axis ($r = 0$) at constant distance $x'' = x - vt$ Lorentz contraction occurs

$$\Phi_{r=0}=\frac{\Phi_o\sqrt{1-v^2/c^2}}{x''}$$

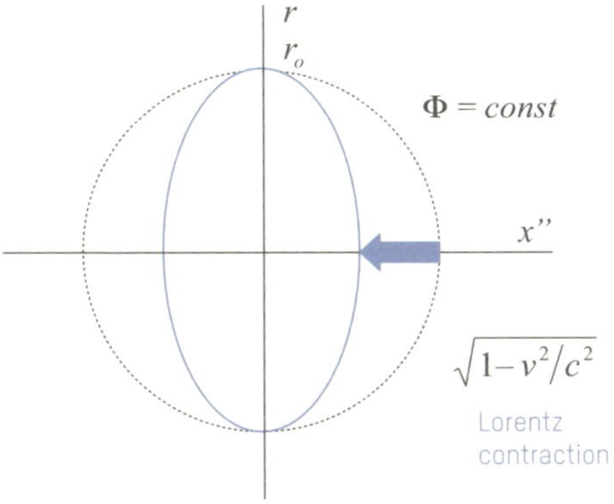

Derivation of inertia from the field energy: $m \cdot dv/dt = F$ is a definition (no law wrestled from experiment), m therein is defined as inertia. The analogous definition of energy is $dE = Fdx$ which can be written as

$$\frac{dE}{dx} = F = \frac{dE}{dv}\frac{dv}{dt}\frac{dt}{dx}$$

where $\frac{dE}{dv}\frac{dt}{dx} = \frac{dE}{dv}\frac{1}{v}$ takes the place of m as resistance against acceleration. It can be calculated from the field energy E_{field} of the influx:

- writing $E_{field} = \Phi \cdot s$, where s represents the mass as sink,
- then inertia is $\frac{d\Phi}{dv}\frac{s}{v}$;
- along and on the x-axis is $\Phi = \frac{\Phi_o \sqrt{1 - v^2/c^2}}{x''}$;
- with derivative $\frac{d\Phi}{dv} = \frac{\Phi_o v}{x''c^2\sqrt{1 - v^2/c^2}}$ comes

$$Inertia = \frac{dE}{dv}\frac{1}{v} = \frac{d\Phi}{dv}\frac{s}{v} = \frac{\Phi_o \cdot s}{x''c^2\sqrt{1 - v^2/c^2}}.$$

If $v = 0$ and x'' represents the inversion radius (whatever its value), then this term corresponds to inertial mass at rest

$$\frac{\Phi_o \cdot s}{x'' c^2} = m$$

and for $v > 0$ results $inertia = \dfrac{m}{\sqrt{1 - v^2/c^2}}$.

Since at the inversion radius $\dfrac{\Phi_o s}{x''}$ corresponds to the entire field energy,

for $v = 0$

$$\frac{E}{c^2} = m \rightarrow E_{rest} = mc^2$$

holds and for $v > 0$

$$E_{field} = m_o c^2 \sqrt{1 - v^2/c^2}$$

which is initially surprising: for the total energy the inverse root term should be expected? Well, to the field energy of the influx the term has to be added representing that the entire dynamics is travelling with velocity v and hence the potential of v^2 has to be attached to it:

$$E = E_{field} + mv^2$$

$$= m_o c^2 \sqrt{1 - v^2/c^2} + \frac{m_o v^2}{\sqrt{1 - v^2/c^2}}$$

$$E = \frac{m_o c^2}{\sqrt{1 - v^2/c^2}}$$

Kinetic energy is the sum of mv^2 and the reduction of the field energy; linearised for clarifying the process:

$$E_{kin} = m_o v^2 - \frac{m_o v^2}{2}$$

Components of the total energy plotted separately:

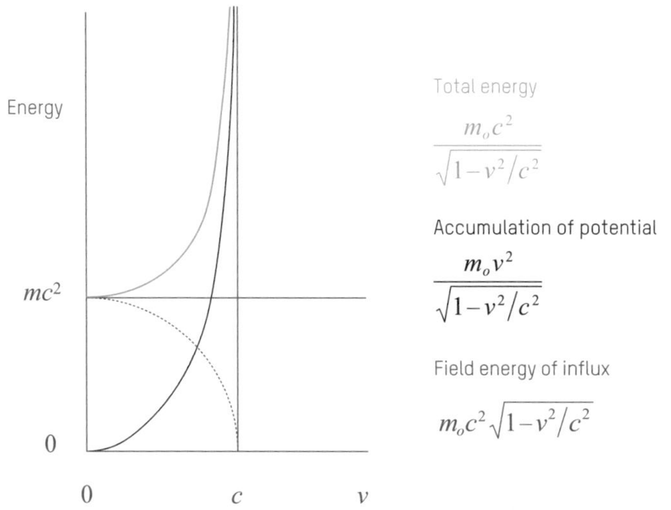

This does in no way resolve what mass is, but only the dynamics is clear, due to which mass offers resistance against acceleration; as well as why this increases with velocity. Deductive physics therefore has to make do with the same measure for mass as inductive physics: a one *kg* mass is that which is precisely as inert as one litre of water.

7. Radiation

The radiation field is the basis for all forces as well as all quantum mechanical phenomena. Its calculations start from the spherical field equation:

$$\Phi_{rr}\left(1-\frac{\Phi_r^2}{c^2}\right)+2\Phi_r\left(\frac{1}{r}-\frac{\Phi_{rt}}{c^2}\right)-\frac{\Phi_{tt}}{c^2}=0$$

For large r is $\Phi_r << c$ and there is an unperturbed spherical wave

$$\Phi=\frac{\Phi_o e^{-i(kr+\omega t)}}{r}$$

Nothing could be won from the solution of the field equation for small radii: the spherical wave $k_r >> 1$ perfectly suffices for representing the "gravitation" phenomenon.

8. Gravitation

Deductive physics explains gravitation analogously to attraction between sink and source as known from gas dynamics. The attractive effect is expressed in gas dynamics by the following formula:

$$F = -\frac{source \cdot sink}{4\pi r^2 \rho}$$

Intuitively it is understandable that two sources repel one another, whereas for two sinks intuitively, but falsely, the opposite—attraction—is expected. The stream lines of their fields, however, are identical like the ones for two sources, only the flow direction is reversed—the interference of the fields is causing repulsion (however, if the sources radiate, they do not interact).

Since gravitation always applies to great distances (in relation to wave length), field compression during approach to the inversion radius does not matter. As in TR the potential can be set $\sim \frac{1}{r}$, specifically $\frac{Gm}{r}$. With G a constant occurs that moots the ignorance of source strength s and density ρ. Mass is both sink and equally strong source—the rest is analogy:

$$E = G\frac{m_1 \cdot m_2}{r} \text{ corresponds to } \frac{1}{4\pi\rho}\frac{sink_1 \cdot sink_2}{r}.$$

From the equation on the left also follows the dimension of G:

$$G = \left[\frac{m^2 kg \cdot m}{s^2 kg^2} = \frac{m^3}{kg \cdot s^2}\right]$$

The radiation field of gravitation is superimposed over the influx; the pressure waves therefore follow the influx stream lines. From the place where the influx velocity reaches c the radiation emanates:

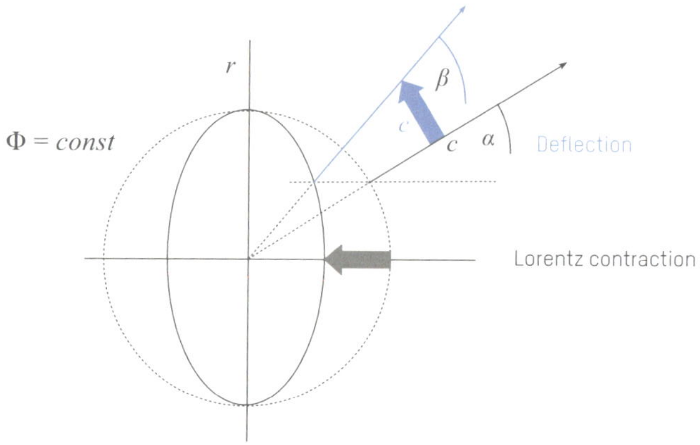

Radiation increase perpendicular to the *x*-axis through deflection is only codetermined by $tg\beta = tg\alpha \Big/ \sqrt{1-v^2/c^2}$; however, for $\alpha = \pi/2$ this relation is exactly the increase. On the *x*-axis deflection is zero. The radiation field thus has the same form as the influx field:

$$\Phi = \frac{\Phi_o}{\sqrt{(x-vt)^2 + r^2\left(1-v^2/c^2\right)}}$$

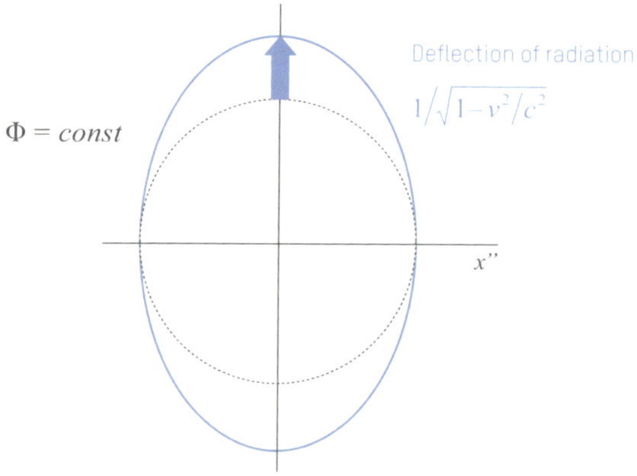

Modification of forces through relative motion: The force in radial direction equals the derivative of the potential with respect to *r*. A companion with velocity *v* observes:

$$F_{radial} = \frac{d\Phi}{dr} = \frac{-\Phi_o}{r^2\sqrt{1-v^2/c^2}} = \frac{F_{unperturbed}}{\sqrt{1-v^2/c^2}}$$

The *axial* force is reduced by the factor $1-v^2/c^2$ because of the counter flux *v*:

$$F_{axial} = F_{unperturbed}\,(1-v^2/c^2)$$

Forces thus increase radially with *v* and decrease axially—as known from the electric field and for the same reason, the deflection by the carrier of the radiation, the influx field:

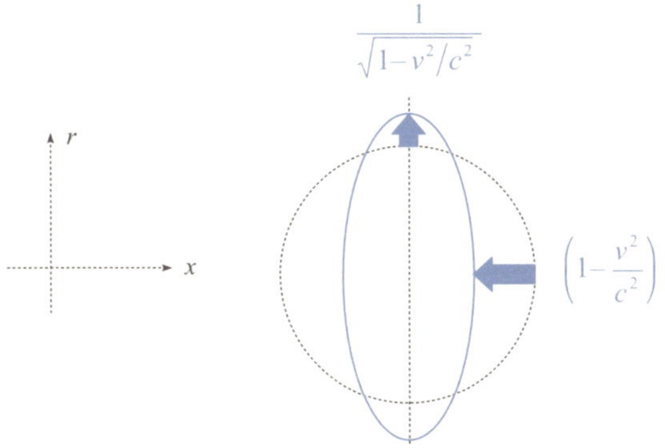

Modification of the equation of motion: Now the momentum theorem can be derived (what inductive physics fails to achieve, but instead in theory of relativity starts from the momentum theorem). To this end both modifications need to be inserted: in direction of *v*

– forces decrease as $\left(1-\frac{v^2}{c^2}\right)$,

– inertia increases as $\frac{1}{\sqrt{1-v^2/c^2}}$

and what at rest is $F = m \cdot dv/dt$ can be written as:

$$F_o\left(1-\frac{v^2}{c^2}\right)=\frac{dv}{dt}\frac{m_o}{\sqrt{1-v^2/c^2}}$$

$$\frac{dv}{dt}\frac{m_o}{(1-v^2/c^2)^{3/2}}=F_o$$

restated

$$\frac{d(mv)}{dt}=F_o$$

Newton's intuition is thus justified retrospectively. Theory of relativity at the core *is* the momentum theorem, everything beyond are ramifications and interpretations. Which the derivation of Einstein's centennial formula from the bare momentum theorem proves:

$$E=\int F_o dx=\int\frac{dx}{dt}\frac{m_o dv}{\left(1-v^2/c^2\right)^{3/2}}=\frac{m_o c^2}{\sqrt{1-v^2/c^2}}$$

with the boundary case $E_{rest} = mc^2$.

9. General Relativity

Equivalence

Einstein postulated for aesthetic reasons that laws should be independent of movement of the frame of reference, which he called equivalence principle. The same laws cogently follow in deductive physics from the dynamics of the continuum: a mass m at distance r from a mass M is passed by the influx towards M like a counter flow with velocity v, and correspondingly for the Lorentz contraction has to be substituted:

$$\frac{v^2}{2} \rightarrow \frac{GM}{r}, \text{ with which comes } m = \frac{m_o}{\sqrt{1 - 2GM/rc^2}}.$$

The Lorentz contraction does not give away whence it originates: from relative velocity or a force field.

This contraction leads to a locally higher frequency $\omega' = \omega/\sqrt{1-2GM/rc^2}$, at the limit to the black hole with $r = 2GM/c^2$. When the light, that is emitted from the surface of a mass M, has overcome the influx, its wave length is relaxed by the factor $(1 - 2GM/rc^2)$, and the frequency that the observer sees is

$$\omega_{observer} = \frac{\omega}{\sqrt{1-2GM/rc^2}}(1-2GM/rc^2)$$

The corresponding quotient is

$$\frac{\omega_{observer} - \omega}{\omega} = \sqrt{1-2GM/rc^2} - 1 \approx -\frac{GM}{rc^2},$$

in the case of a light source on the surface of the sun: $2 \cdot 10^{-6}$, the observed red shift.

Radial Equation of Theory of Relativity (TR)

The predictions of the so called radial equation of the general theory of relativity used for calculating perihelion shift and light deflection aided the breakthrough of general relativity at the 1919 eclipse. It is nothing but the formula $E_{kin} + E_{pot} = const$ modified by Lorentz contractions, which becomes apparent, if it is linearised (for sake of simplicity without angular momentum): In TR notation with r = distance from centre of rotation, $A^{-1} = B = 1 - 2GM/rc^2$ it reads

$$\frac{A}{B^2}\dot{r}^2 - \frac{c^2}{B} = -\,const$$

With the approximation $A^{-1} = B = 1$ for the term with $\dot{r}^2$, as well as $B^{-1} \approx 1 + 2GM/rc^2$ for the term with c^2 (because of its magnitude) and finally $const = c^2$ Newton surfaces non-relativistically:

$$\dot{r}^2 - c^2\left(1 + \frac{2GM}{rc^2}\right) = -\,const$$

$$\frac{\dot{r}^2}{2} - \frac{GM}{r} = 0$$

From there it can now be reconstructed based on the modifications of force and inertia caused by the Lorentz contraction:

$$-\,E_{kin} \qquad \rightarrow \frac{r^2}{2}\frac{1}{\sqrt{1-\dot{r}^2/c^2}}\frac{1}{\sqrt{1-2GM/c^2}}$$

$$-\,E_{pot} \qquad \rightarrow -\frac{GM}{r}\left(1-\dot{r}^2/c^2\right)\left(1-2GM/c^2\right)$$

For $\dot{r} << c$ and no other forces, which holds throughout the universe, the potentials can be set equal $\dot{r}^2/2 = GM/r$ and the modifications of the potentials are written thus:

$$-\,E_{pot} \qquad \rightarrow -\frac{GM}{r}B^2$$

$$-\,E_{kin} \qquad \rightarrow \frac{\dot{r}^2}{2}A$$

Inserting this into the non-relativistic original equation, dividing it by B^2 and amending on both sides by $-c^2$ again provides the radial equation

$$\frac{\dot{r}^2}{2}A - \frac{GM}{r}B^2 = 0$$

$$\dot{r}^2\frac{A}{B^2} - \frac{2GM}{r} - c^2 = -c^2$$

$$\dot{r}^2\frac{A}{B^2} - \frac{c^2}{B} = -c^2$$

It derives from the linearisation of the momentum theorem (because of this linearisation the Robertson-Walker metric applies only to weak fields).

Lorentz Invariance

That laws of nature should be independent from state of motion can be fulfilled from certain perspectives—but in the end the idea is not tenable, as even TR itself proves:

- At rest inert and heavy mass are indeed "equal" (see below what "equal" means),
- but at velocity v inertia increases in the direction of v by $(1-v^2/c^2)^{-1/2}$,
- whereas gravitation and Coulomb forces decrease in the same direction by $(1-v^2/c^2)$,
- whence follows that there can be no transformation neutralising both changes simultaneously.

Einstein: "The identity of [...] heavy mass and of inert mass is experienced fact established with maximum precision (Eötvös experiment), for which classical mechanics [has] no explanation". Science would only do it justice, "once it has reduced that numerical identity [sic!] to identity in essence".

What does he mean by "equal"? Numerical identity is built in via G:
- what offers resistance against acceleration of m (at rest $m \cdot dv/dt$),
- that also has an attractive potential, measured in experiment, at distance r of $m \cdot G/r$.

The nature of both is entirely different; necessarily equal are amounts of continuum turned over in influx and radiation.

10. Cosmology

The expansion of the universe is powered

- in *deductive physics* by the pressure of compressed continuum (masses forming in it after the "primordial compression" are "floating" along in the continuum),
- in the *Standard model of cosmology* by the energy of compressed matter whose density and temperature tend towards infinity.

Therefore there are results differing from TR, albeit formally similar: in deductive physics

a. the Hubble constant refers to flight of continuum—not to flight of masses;
b. the Einstein-de Sitter universe (Standard Model) cannot exist due to the inexceedability of the speed of light;
c. cosmological red shift presents itself similar to the standard model, but for different reasons;
d. there is no need for any "dark matter";
e. the "cosmological constant" currently being revitalised turns out to be unsuitable for explaining the accelerated expansion of the universe,
f. there is no need for any "dark energy";
g. values of G, $\hbar$ and inertia change with expansion;
h. there are more degrees of freedom for the development of the universe than in TR.

This will be justified in the following:

a. Hubble Constant

As to be seen for the influx towards a black hole the density distribution beyond a few Schwarzschild radii is almost constant. Analogously it is practically constant from big bang to the event horizon. This means $\partial\rho/\partial r = 0$ for the continuity and Euler equations ("flat universe"—mandatory) whereby they reduce to

$$C \quad \frac{\partial \dot{r}}{\partial r} + \frac{2\dot{r}}{r} + \frac{1}{\rho}\frac{\partial \rho}{\partial t} = 0$$

$$M \quad \frac{\partial \dot{r}}{\partial t} + \dot{r}\frac{\partial \dot{r}}{\partial r} = 0$$

From C follows $\rho \sim t^{-3}$; from M follows $\frac{d\dot{r}}{dt} = 0$, which means: no radial acceleration. The event horizon always moves away with c, and every point in between proportionally with $\dot{r} = \frac{r}{r_{horizon}} c$.

With this for each radius $r = r(t_o) + \dot{r}t$ holds, from which the Hubble constant results:

$$H = \frac{\dot{r}}{r} = \frac{1}{r(t_o)/\dot{r} + t} \approx \frac{1}{t}$$

The formula corresponds to the one from TR, the difference being that $\dot{r}$ in deductive physics refers to the continuum, in TR on the other hand to the masses.

b. Einstein-de Sitter Universe

TR bases derivation of the world models on its field equation. It is linearised through the Robertson-Walker metric (Friedmann models). A particular one of them is called Einstein-de Sitter universe, serves as Standard Model for cosmology and leads to untenable singularities for $t \to 0$.

To prove this deductive physics integrates the momentum theorem:

$$F\left(1 - \frac{v^2}{c^2}\right) = \frac{dv}{dt}\frac{m}{\sqrt{1 - v^2/c^2}}$$

If the force F stems from the gravitation of the mass M contained in the universe of radius R, then

$$\frac{GMm}{R} = mc^2\left(\frac{1}{\sqrt{1-\dot{R}^2/c^2}}-1\right)$$

Transformed for comparability with Friedmann models, as well as with $R_S = 2GM/c^2$ substituted,

$$\frac{\dot{R}^2}{2}-\frac{GM}{R}\cdot\frac{1+R_s/4R}{(1+R_s/2R)^2}=0 \text{ results.}$$

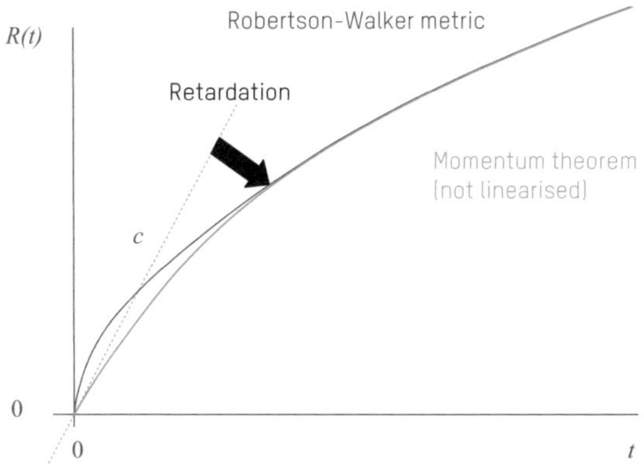

For large R the factor accompanying GM/R vanishes, and the Einstein-de Sitter universe (Friedmann model with $k, \Lambda = 0$) is at hand:

$$\frac{\dot{R}^2}{2}-\frac{GM}{R}=0$$

For $R = 0$,

- in the Einstein-de Sitter universe tends $\dot{R} \rightarrow \infty$ and the total potential is infinity minus infinity,
- in deductive physics $\dot{R} = c$ and the potential is always

$$\frac{c^2}{\sqrt{1-\dot{R}^2/c^2}}-\frac{GM}{R}=c^2,$$

which is why deductive physics does not need an "inflation": with $\dot{R}=c$ and $R=0$ there is an end to it, all material structures disappear—and the model has to yield to elementary particle physics.

c. Cosmological Red Shift

The following three causes of red shift are superimposed onto each other:

(a) The first one is based on the flight velocity $\dot{R}$ and is linear, since the influx towards the light emitting masses is not being contracted (masses initially at rest in fleeing continuum):

$$z_{flight} = \frac{\omega_{flight}}{\omega} - 1 = \frac{c - \dot{R}}{c} - 1 = \frac{\dot{R}}{c}$$

(b) Masses on a sphere with radius R are retarded in their expansion by the gravity of the enclosed masses M, whence a Lorentz contraction and hence gravitational red shift z_{kin} result

$$z_{grav} = \frac{\omega_{grav}}{\omega} - 1 = \frac{1}{\sqrt{1 - 2GM/Rc^2}} - 1 \cong \frac{GM}{Rc^2}$$

(c) This gravitational work reduces $\dot{R}$. Here $\frac{\Delta\dot{R}^2}{2} = \frac{GM}{R}$, from which

$$z_{kin} = -\frac{\Delta\dot{R}}{c} = -\sqrt{\frac{2GM}{Rc^2}}$$

Total red shift is

$$z_{total} = z_{flight} + z_{grav} + z_{kin}$$

$$z_{total} = \frac{\dot{R}}{c} + \frac{GM}{Rc^2} - \sqrt{\frac{2GM}{Rc^2}}$$

The formula of TR (with D, "physical distance", H Hubble constant)

$$z = \frac{HD}{c} + \frac{\left(1 - \ddot{R}R/\dot{R}^2\right)}{2c^2} H^2 D^2$$

made comparable by $R = D$, $\frac{HD}{c} = \frac{\dot{R}}{c}, \frac{\dot{R}^2}{2} = \frac{GM}{R}$, as well as $\ddot{R} = -\frac{GM}{R^2}$

leads to $z = \frac{\dot{R}}{c} + \frac{3}{4}\frac{GM}{Rc^2}$.

The difference between deductive physics and TR is that

- in TR $\dot{R}$ stands for the flight velocity and R for the location of the masses,
- while in deductive physics $\dot{R} - \sqrt{\frac{2GM}{R}}$ is the flight velocity of the masses and R stands for the continuum.

d. Dark Matter

The masses of galaxies do not suffice for explaining their dynamics—it would take five to six times more of them. Inductive physics makes do with "dark matter". With the term "matter" is implied: *attraction* = $G \cdot$ *matter*.

In deductive physics the black hole provides the wanted attraction, however, not by gravitation, but by influx: the masses entering the maelstrom of a black hole "float" along. A corresponding force can only be formulated negatively: as that which would be necessary to keep the masses back.

e. Cosmological Constant

For explaining the measured accelerated red shift, cosmology is again thinking of Einstein's cosmological constant. For realising what TR means by this deductive physics sets the mass density to zero in the Friedmann models,

$$\dot{R}^2 + kc^2 = \frac{GM}{R} + \frac{\Lambda c^2}{3} R^2$$

likewise k (for Einstein-de Sitter), which leads to the following contradiction:

$$\frac{\dot{R}^2}{R^2} = \frac{\Lambda c^2}{3} \approx \frac{1}{t^2} \neq const$$

and implies that the cosmological constant is not suitable for explaining the accelerated expansion.

f. Dark Energy

Deductive physics discovers another cause for the unexplained acceleration of the universe's expansion than "dark energy": the 10^{10} black holes of our universe essentially form a cosmological gas, whose constituents repel each other like sinks. They act on the universe as a whole quite contrary to their effect within the galaxy.

This "dark energy", E_{dark}, can be calculated as work of the black holes against each other. As the dark matter, M_{dark}, is treated as matter by inductive physics, which provides the otherwise unexplained attraction within the galaxies, their dispersing force can be computed as the inverse of this attraction:

$$dE_{dark} = \frac{GM_{dark}}{R} dM_{dark}$$

$$E_{dark} = \frac{GM_{dark}^2}{R} \frac{3}{5}$$

The equivalent mass M_{aequiv} (with visible mass M, R event horizon)

$$M_{aequiv} = \frac{E_{dark}}{c^2} = \frac{2GM}{c^2 R} \frac{3}{2 \cdot 5} \frac{M_{dark}^2}{M}$$

$$\frac{M_{aequiv}}{M_{dark}} = 0.3 \frac{R_{Schwarzschild}}{R_{real}} \frac{M_{dark}}{M}$$

With the current estimate of $M_{dark} / M = 5.75$ and with $R_{Schwarzschild} / R = 1.84$, the energy distribution would correspond to the current values of 4% visible, 23% dark matter and 73% dark energy. However, there is insufficient evidence for the assumed ratio of $R_{Schwarzschild} / R$.

The black holes do not accelerate the expansion of the continuum, but of the galaxies—5.75^2 faster than gravity retards them. Fusion of black holes is impossible in deductive physics because of their mutual repulsion.

g. Permanence of the Constants

In deductive physics only the model assumption of a continuum with potential c^2 stays permanent. On the other hand gravitational constant G and action quantum h are not model assumptions but measurements; they vary with the event horizon in the same way that elementary masses vary in their inertia and attractive force:

		Development over time
Planck's constant	correlated with free path length	↑
Gravitational constant	with density of continuum in the denominator	↑
Inertia	due to decreasing influx strength	↓
Gravitation	proportional to influx	↓

h. Cosmological Degrees of Freedom

The dynamics of the universe offers sufficiently many degrees of freedom to avoid mysterious models: for example

- the expansion of the universe is vied for by the black holes within each galaxy which absorb matter and continuum (the respective event horizon being limited by the speed of light);
- opposite dynamics of continuum and masses is possible.

II – Quantum Mechanics

In deductive physics that which is subject matter of quantum mechanics in inductive physics arises cogently from mass dynamics:

- at the root lies the radiation of mass dynamics,
- radiations are being modified at relative velocities and within force fields and interfere with original radiations (at force free rest quantum mechanical phenomena do not occur),
- the frequencies of the interference wave resonate with mechanical movements and yield eigenvalues,
- with interference and resonance field energy is being reduced which changes to zero point motion in certain geometries.

While inductive physics posits probability calculations as proper laws of nature for which it cannot provide any causes, deductive physics derives all phenomena and laws from three causes underpinning the phenomena:

The treatment of the quantum mechanical wave as an interference wave, whose envelope runs at $v/2$, if the enveloped wave packet has velocity v establishes:

Envelope at $\nu/2$ (interacts with other particles)

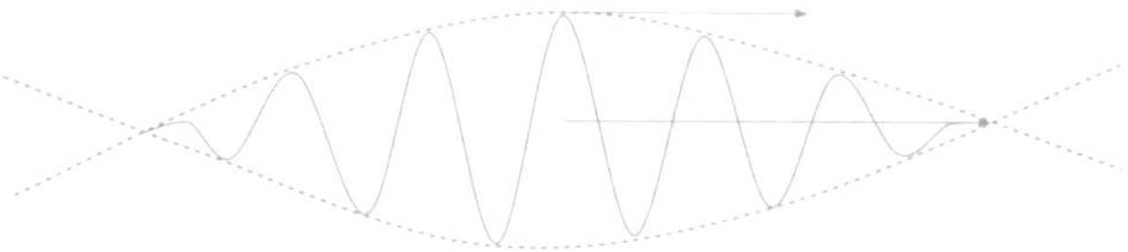

Wave packet at ν (represents the locations of the particle)

1. the hermiticity of the quantum mechanical wave,
2. the de Broglie-Einstein relations,
3. the rest frequency of a mass (Dirac[1928]),
4. scattering matrices,
5. the Heisenberg uncertainty relation;

the resonance of mechanical and quantum mechanical frequency establishes

6. the action quantum,
7. the properties of the hydrogen atom,
8. the properties of the harmonic oscillator,
9. the Schrödinger equation;

the annihilation of field energy establishes

10. the zero point motion: conversion into kinetic energy from radiation energy being cancelled by interference.

1. Hermiticity of the QM Wave

The wave function ψ has to be hermitian for the calculations of QM to come out even, something inductive physics cannot explain any further. Deductive physics explains it as follows:

The wave equations for velocity v and density ρ are satisfied by arbitrary waves $e^{i(kx - \omega t)}$. However, both waves have to satisfy the continuity equation C and the Euler equation M as real functions, for which they would need to be synchronous, e.g. v, $I \sim \cos$, and the absolute amplitudes $v_o = cI_o$. Synchronous solutions do not satisfy the condition of the potential being constant $v^2 + c^2I^2 = const$, and if $v \sim \cos$, then $I \sim \sin$ which presupposes phase shift.

Only hermitian functions satisfy all conditions: Inserting for an interference wave, arising for instance from the superposition of $v_o e^{i(kx-\omega t)}$ + $v_o e^{i(kx+\omega t)}$, into C or M gives a function for I

$$v = v_o \cos \omega t \cdot (\cos kx + i \sin kx)$$
$$I = I_o \sin \omega t \cdot (\sin kx - i \cos kx)$$

which satisfies both wave equations as well as the constancy of the potential (with $v_o = cI_o$; $\omega = kc$; $v^2 \rightarrow vv^*$; $I^2 \rightarrow II^*$)

$$vv^* = c^2II^* = v_o^2 \cos^2 \omega t + c^2 I_o^2 \sin^2 \omega t = v_o^2 = const$$

v and I are distinct expressions of a single dynamics that has to be described by two variables. The imaginary part of one variable represents the formally suppressed other variable.

2. de Broglie-Einstein Relations

At the outset of the de Broglie-Einstein relations, Lorentz contractions result in changes to frequencies and wave lengths of radiations of the mass dynamics:

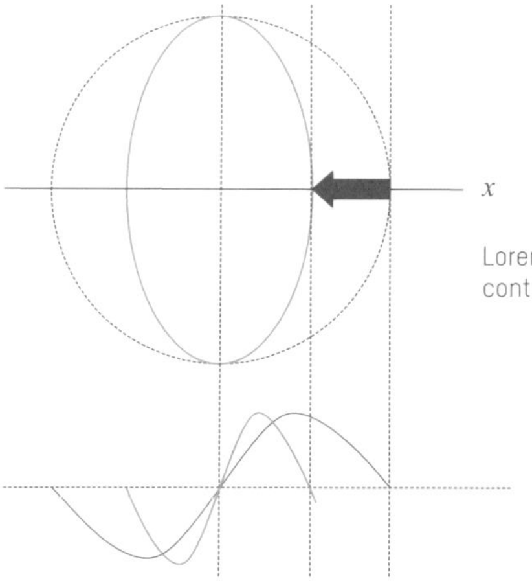

Lorentz contractions are caused by relative velocities and potential fields:

The coordinates *x'* and *t'* associated with a mass *m* moving with *v* relative to an observer change with *v* versus *x* and *t* of the observer:

$$x' = \frac{x - vt}{\sqrt{1 - v^2/c^2}} \text{ as well as } t' = \frac{t - vx/c^2}{\sqrt{1 - v^2/c^2}}$$

With the item at hand being radiation, $x = ct$:

$$x' = x\frac{1 - v/c}{\sqrt{1 - v^2/c^2}} \text{ as well as } t' = t\frac{1 - v/c}{\sqrt{1 - v^2/c^2}}$$

Frequency ω and wave number k change correspondingly. For $\omega t' = \omega' t$ and $kx' = k'x$

$$\omega' = \omega \frac{1 - v/c}{\sqrt{1 - v^2/c^2}} \text{ as well as } k' = k \frac{1 - v/c}{\sqrt{1 - v^2/c^2}}$$

The gravitational field causes similar modifications to frequency and wave number. Instead of the opposite flux with v the influx towards a mass m is acting and the kinetic term in the Lorentz contraction has to be replaced with the gravitational potential: $v^2/2 \rightarrow Gm/r$

$$\omega' = \omega \frac{1}{\sqrt{1 - 2Gm/rc^2}} \text{ as well as } k' = k \frac{1}{\sqrt{1 - 2Gm/rc^2}}$$

This yields the gravitational red shift at the surface of a star with mass m and radius r.

The Lorentz contraction leads to an increased frequency $\omega' = \omega/\sqrt{1 - 2Gm/rc^2}$. When the light from the surface of a mass has surmounted the influx that has contracted the field, $(1{-}2Gm/rc^2)$, the wave length is relaxed and the frequency that an observer sees is

$$\omega_{observer} = \frac{\omega}{\sqrt{1 - 2Gm/rc^2}}(1{-}2Gm/rc^2)$$

The resulting red shift is

$$z_{grav} = \frac{\omega' - \omega}{\omega} = \frac{1}{\sqrt{1 - 2Gm/rc^2}} - 1 \approx \frac{Gm}{rc^2}$$

In the case of light from the surface of the sun, $-2 \cdot 10^{-6}$, the value measured in 1919.

Within the Coulomb potential of an electron results a red shift corresponding to the fine structure constant α:

$$\omega' = \omega \frac{1}{\sqrt{1 - 2e^2/mc^2 r}}$$

$$z_{Coulomb} = \frac{\omega' - \omega}{\omega} \approx \frac{e^2}{mc^2 r}$$

at $r = \lambda\!\!\!{}^{-}_e : z_{Coulomb} \approx \dfrac{e^2 \cdot mc}{mc^2 \cdot \hbar} = \alpha$

The de Broglie-Einstein relations derive purely from the treatment of the quantum mechanical wave as an interference wave:

- in general for interference waves: $\dfrac{\omega}{k} = \dfrac{v}{2}$ as well as $\dfrac{\partial\omega}{\partial k} = v$ hold,
- hence $\dfrac{\partial\omega}{\partial k} = \dfrac{2\omega}{k} \rightarrow \omega = k^2 const \rightarrow k \sim v;\ \omega \sim \dfrac{v^2}{2}$,
- with $const = \dfrac{m}{\hbar}$ from experiment result

  $\omega = \dfrac{mv^2}{2\hbar}$ as well as $k = \dfrac{mv}{\hbar}$.

These turn out to be linearisations of the derivation which now follows and which deductive physics needs for elementary particles. This requires the above modifications of frequency and wave number: If a radiation of frequency ω at the place of emission arrives with ω' at the observer, and the emitted wave interferes with the reflected one,

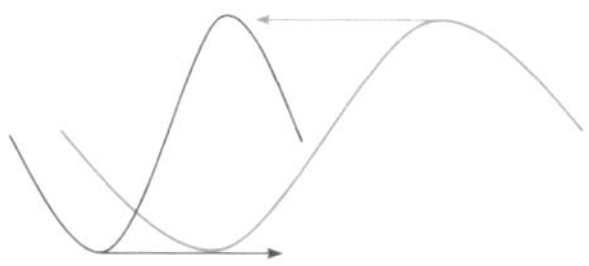

then the amplitude of the interference wave is subject to the trigonometric relation

$$\sin\omega + \sin\omega' = 2\sin\frac{\omega' + \omega}{2} \cdot \cos\frac{\omega' - \omega}{2}$$

Only the envelope cosine is relevant for interactions. Its frequency ω and wave number k are computed from the unknown quantities at the place of emission ω_u, k_u

$$\omega = \frac{\omega_u}{2}\left(\frac{1+v/c}{\sqrt{1-v^2/c^2}} - 1\right)$$

$$k = \frac{k_u}{2}\left(\frac{1+v/c}{\sqrt{1-v^2/c^2}} + 1\right)$$

The plus sign for k results from phase velocity inversion due to reflection. The phase velocity ω/k is, with $\omega_u/k_u = c$,

$$\frac{\omega}{k} = \frac{c^2}{v}\left(1 - \sqrt{1-v^2/c^2}\right)$$

For determining ω, k individually a further condition is required. It is contained in the group velocity $\partial\omega/\partial k = v$. After inserting and transforming follows

$$\frac{dk}{k} = \frac{dv}{v}\frac{1}{1-v^2/c^2}$$

hence by integration

$$k = \frac{v}{\sqrt{1-v^2/c^2}}\, const$$

The constant results from experiment and is $\frac{m}{\hbar}$. With this, after inserting into ω/k,

$$\omega = \frac{mc^2}{\hbar}\left(\frac{1}{\sqrt{1-v^2/c^2}} - 1\right) \text{ and } k = \frac{mv}{\hbar}\frac{1}{\sqrt{1-v^2/c^2}}$$

The usual de Broglie-Einstein relations are obtained via linearisation of fractions and root terms as well as neglecting third powers of v, from which ensue

$$\omega = \frac{mv^2}{2\hbar} \text{ as well as } k = \frac{mv}{\hbar}.$$

Since the particle generates the wave packet, the velocity of the particle is identical with the velocity of the wave packet, and the nature of the interference wave is confirmed in these linearisations, too

$$v_{envelope} = \frac{\omega}{k} = \frac{mv^2}{2\hbar}\frac{\hbar}{mv} = \frac{v}{2}$$

$$v_{wave\ packet} = \frac{\partial\omega}{\partial k} = \frac{\partial\omega}{\partial v}\frac{\partial v}{\partial k} = \frac{mv}{\hbar}\frac{\hbar}{m} = v$$

3. Rest Frequency

If ω is interpreted as $\frac{1}{2}(\omega_{rest}' - \omega_{rest})$, then a rest frequency results

$$\omega = \frac{mv^2}{2\hbar} = \frac{\omega_{rest}}{2}\left(\frac{1}{\sqrt{1-v^2/c^2}} - 1\right)$$

$$\omega_{rest} = \frac{2mc^2}{\hbar}$$

As the radiation of a mass at rest has phase velocity c,

$$k_{rest} = \frac{\omega_{rest}}{c} = \frac{2mc}{\hbar}$$

holds from which derives an angular momentum $\omega_{rest}\, ƛ_{rest}^2\, m = \hbar/2$ which matches fermion spin.

The constant $\frac{m}{\hbar}$ that the experiment yielded for the de Broglie-Einstein relations reveals its origin in resonances

$$\frac{m}{\hbar} = \frac{mc}{\hbar}\frac{1}{c} = \frac{1}{ƛc} \rightarrow mƛc = \hbar.$$

4. Scattering Matrices

Resonance governs scattering processes. Scattering of the wave of a mass m with velocity v by a box potential U in deductive physics means: the wave comes up onto a local flux with a virtual velocity $v_{virt} = \sqrt{2U/m}$ in the direction of v. The relative velocity of the mass against the continuum is reduced and wave length is increased inversely proportional to the relative velocity.

In the chart the mass is coming from the left with $v \rightarrow \lambda\!\!\!^{-} = \hbar/mv$. Its wave is partially reflected at the box, the remainder exits on the right. Inside the box $\lambda\!\!\!^{-}$ increases inversely proportional to $\sqrt{1-2U/mv^2}$ and afterwards again takes its original length. Without display of imaginary part, unperturbed inbound wave as thin line:

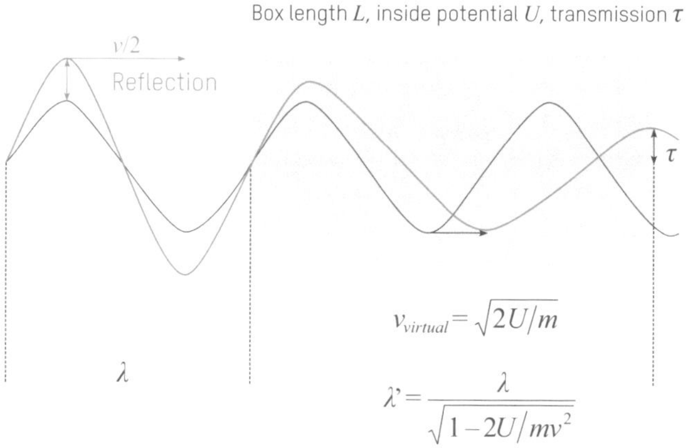

The wave amplitudes and their gradients are continuous at entry to and exit from the box which leads to four equations as well as the solution for the coefficient of reflection R (for $E > U$, formula from QM text book, rewritten)

$$R = \frac{1}{1+\zeta^2} \qquad \text{with } \zeta = \frac{v\hbar}{U\bar{\lambda}\cdot \sin L/\bar{\lambda}}$$

R is zero for $L = \bar{\lambda}\, n\pi = n\frac{\lambda}{2}$ and a maximum for $L = \left(n + \frac{1}{2}\right)\frac{\lambda}{2}$

This result for $n = 0$ demonstrates that scattering is governed by resonance:

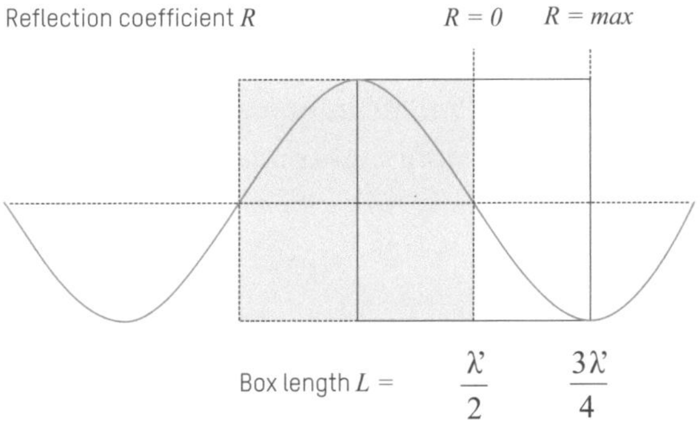

5. Uncertainty

The Heisenberg uncertainty relation, according to which the uncertainty of cognizance of the locus of a mass Δq multiplied with the uncertainty of its momentum Δp always equals at least the Planck constant: $\Delta q \cdot \Delta p \geq \hbar$, is explained as follows:

- since the particle generates the wave, it is a priori located *within* the wave train with maximal amplitude,
- however, it cannot be inferred, *where* in the wave train and "spatial uncertainty" corresponds to one wave length, $\Delta q = \lambdabar$
- the wave length is inversely proportional to momentum and that consists only of "momentum uncertainty", thus $\lambdabar = \hbar/\Delta p$,
- hence "spatial uncertainty times momentum uncertainty" yields

$$\frac{\hbar}{\Delta p}\Delta p = \hbar.$$

The "lack of knowledge" is therefore subjective: the particle does have an exact location, but does not communicate it.

6. Action Quantum

"Planck's quantum of action" from 1900, the base of quantum theory, contains the term "action", because it has dimension of energy multiplied by time, "quantum", because this action only occurs in integral multiples of $\hbar$. By dimension it is also an angular momentum. In deductive physics this angular momentum results in general from synchronisation of mechanical frequencies with those of interference waves. The angular momentum l of a mass m orbiting a point with frequency ω at constant distance r is $l = m \cdot \omega \cdot r^2$.

Multiplying both dividend and divisor with ω yields

$$l = m \cdot \omega \cdot r^2 = \frac{m\omega^2 r^2}{\omega} = \frac{E_{pot}}{\omega}$$

Inserting

- the synchronisation condition $\omega_{QM} = \omega \cdot n$,
- the de Broglie-Einstein relation $\omega_{QM} = E_{pot}/\hbar$,

always yields the same value $n\hbar$ independent of mass and frequency for the angular momentum:

$$l = \frac{E_{pot}}{E_{pot}/\hbar n} = n\hbar$$

This underlines:

- the action quantum is no quantum which acts, but it is the angular momentum which results in general from the synchronisation of mechanical and quantum mechanical frequency—albeit not above atomic dimensions: the angular momentum of a railroad engine is not $\hbar$,
- $\hbar$ has to be understood as a property of continuum alone (it correlates with continuum's free path length).

If the distance of m to the centre of rotation, r, is not fixed, but determined by a force dE/dr, whose value changes with r, the equilibrium of forces is $m\omega^2 r = dE/dr$. From this results

$$l = m\omega r^2 = \sqrt{m\frac{dE}{dr}r^3}$$

Two examples show how this simple formula leads to the known results:

Hydrogen atom $\quad \frac{dE}{dr} = \frac{e^2}{r^2} \rightarrow l = \sqrt{m_e \frac{e^2}{r^2} r^3}$

resonance at $\quad l = n\hbar$

resulting radius $\quad r_n = n^2 \frac{\hbar^2}{m_e e^2} = n^2 r_{Bohr}$

Harmonic oscillator $\quad \frac{dE}{dx} = ax \rightarrow l = \sqrt{m \cdot ax \cdot x^3} = m\omega x^2$

resonance at $\quad l = n\hbar$

resulting displacement $\quad m\omega x_n^2 = n\hbar \rightarrow x_n = \sqrt{\frac{n\hbar}{m\omega}}$

7. Hydrogen Atom

Computation of the radius: The hydrogen atom has to be treated as an oscillator. The assumption $m_e \frac{d^2r}{dt^2} = -\frac{e^2}{r^2}$ has to be extended relativistically due to the otherwise exceeded speed of light for $r \to 0$ (for the sake of simplicity without angular momentum):

$$\frac{m_e}{\sqrt{1-\dot{r}^2/c^2}} \frac{d^2r}{dt^2} = -\frac{e^2}{r^2}\left(1-\frac{\dot{r}^2}{c^2}\right)$$

Integration over t from an displacement r_o where $\dot{r} = 0$ yields

$$m_e c^2\left(\frac{1}{\sqrt{1-\dot{r}^2/c^2}}-1\right) = e^2\left(\frac{1}{r}-\frac{1}{r_o}\right)$$

(mechanically computed $\dot{r}$ would reach speed of light c for $r = 0$: this would be an illeg itimate extrapolation into quantum mechanical terrain).

For minimising energy one has to synchronise

- interference frequency $\omega_{QM} = E_{pot}/\hbar = e^2/\hbar r_o$,
- mechanical frequency (oscillation) from $\omega^2_{mech} r_o m = e^2/r_o^2$, $\omega_{mech} = \sqrt{e^2/mr_o^3}$,
- $\omega_{QM} = n \cdot \omega_{mech} \rightarrow \frac{e^2}{\hbar r_o} = n\sqrt{\frac{e^2}{mr_o^3}}$,
- $r_n = n^2 \cdot \frac{\hbar^2}{e^2 m} = n^2 r_{Bohr}$.

Calculation of energy: The amplitude of the electric field in deductive physics has the dimension $\sqrt{force}$ and is $\frac{e}{r}$ (hence $E_{pot} = \int_\infty^{r_o} e^2/r^2 dr = -\frac{e^2}{r_o}$).

Correspondingly the absolute amplitude of the interference wave is $\psi_o = \frac{e}{r_o}$. The energy of the interference wave is the integral $\int_0^\infty \psi^2 dr$.

For $n = 1$ and with ψ from the solution of the Schrödinger equation (it is derived on p. 76ff.)

$$E_{interference\ wave} = \psi_o^2 \int_0^\infty e^{-2r/r_{Bohr}} dr = \psi_o^2 \frac{r_{Bohr}}{2} = +\frac{e^2}{2r_{Bohr}}$$

For $n > 1$ a factor $1/n^2$ enters, with which the total energy turns out to be Riedberg energy

$$E_{total} = -\frac{e^2}{n^2 r_{Bohr}} + \frac{e^2}{2n^2 r_{Bohr}} = -\frac{e^2}{2n^2 r_{Bohr}}$$

Consequently interference energy reduces potential energy by one half via annihilation of Coulomb waves.

Inserting total energy $E_{total} = -\frac{e^2}{n^2 2r_{Bohr}}$ into the Schrödinger equation

$$-\frac{e^2}{n^2 2r_{Bohr}} = -\frac{\Delta\psi}{\psi}\frac{\hbar^2}{2m} - \frac{e^2}{r}$$

$$\frac{\Delta\psi}{\psi} = \frac{1}{n^2 r_{Bohr}^2} - \frac{2}{r_{Bohr} r}$$

yields the solution $\psi_1 = \frac{e}{r_{Bohr}} e^{-r/r_{Bohr}}$ for $n = 1$ (as above used without derivation).

The interference energy has a minimum and a maximum of $e^2/2n^2 r_{Bohr}$ above or below the attraction energy between n and $n + 1$ for $r = n^2 r_{Bohr}$. Qualitative profile of total energy:

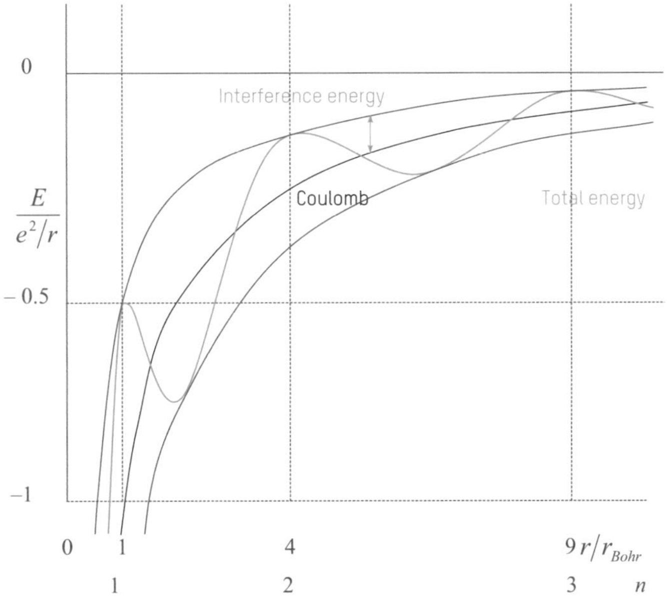

This explanation of the hydrogen atom highlights why the electron being accelerated back and forth fails to radiate: because its radiation enters into the interference wave whose energy settles into a constant value and turns stationary.

8. Harmonic Oscillator

From $m\frac{d^2x}{dt^2} = -a \cdot x$ derives the mechanical frequency $\omega^2_{mech} = \frac{a}{m}$

Synchronisation requires that it adds up n times to quantum mechanical frequency, from which result

$$\omega_{QM} = \frac{E_{kin}}{\hbar} = \frac{m\omega^2_{mech}x^2_n}{2\hbar} = n\omega_{mech}$$

$$x^2_n = \frac{2n\hbar}{\sqrt{am}}$$

as well as $E_n = \frac{2an\hbar}{2\sqrt{am}} = n\hbar\omega.$

On top of this energy comes the zero-point energy: field energy annihilated by the interference wave transitions into energy of an oscillation. It computes as follows:

- phase velocity of the interference wave is $v_{phase} = \frac{\omega}{k}$,
- the phase is standing still and the mass m is passing it with velocity $v_o = \omega x_o$ at $x = 0$,
- set equal as relative velocities $v_o = \frac{\omega}{k} = \omega x_o$, follows $x_o = \frac{1}{k} = \lambda\!\!\!^{-}_o$.

The equality of the values of absolute kinetic and potential energy

$\frac{ax^2_o}{2} = \frac{p^2_o}{2m} = \frac{\hbar^2}{2m\lambda\!\!\!^{-2}_o} = \frac{\hbar^2}{2mx^2_o}$ leads to $x^2_o = \frac{\hbar}{\sqrt{am}}$ as well as

$$E_o = \frac{\hbar a}{2\sqrt{am}} = \frac{\hbar\omega}{2}.$$

The total energy is the sum of kinetic and zero-point energy plotted as a function of displacement:

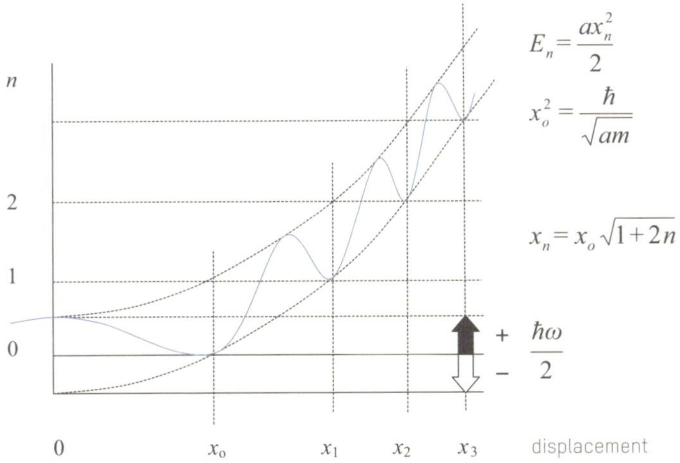

Minima represent the established eigenvalues; locations in between are not occupied spontaneously and count as "forbidden":

$E(x_o)$, the total energy at x_o vanishes, if the negative interference energy of zero-point motion is included, which QM, however, fails to do and obtains a positive energy of $\hbar\omega/2$—but cannot identify where it originates.

The three energy terms of a zero-point motion presented synoptically, sum zero:

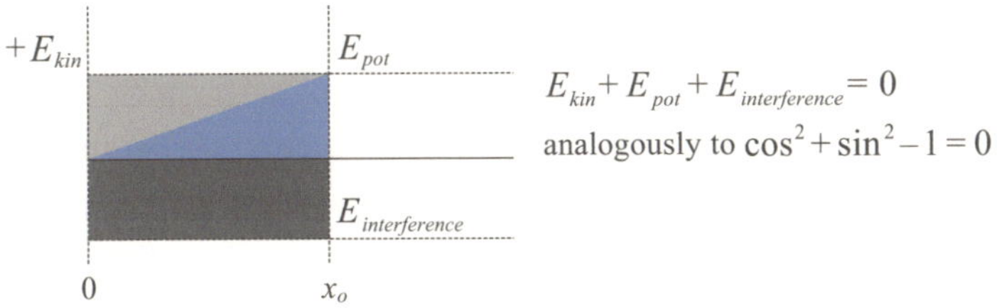

The Schrödinger equation with zero-point energy inserted yields the interference wave

$$\frac{\hbar\omega}{2} = \frac{\hbar^2 \psi''}{2m\psi} + \frac{ax^2}{2}$$

$$\frac{\psi''}{\psi} = \frac{1}{x_o^2} - \frac{x^2}{x_o^4}$$

$$\psi = \psi_o e^{-x^2/2x_o^2}$$

It behaves like a cosine forced to coast down with the same curvature at $x = 0$:

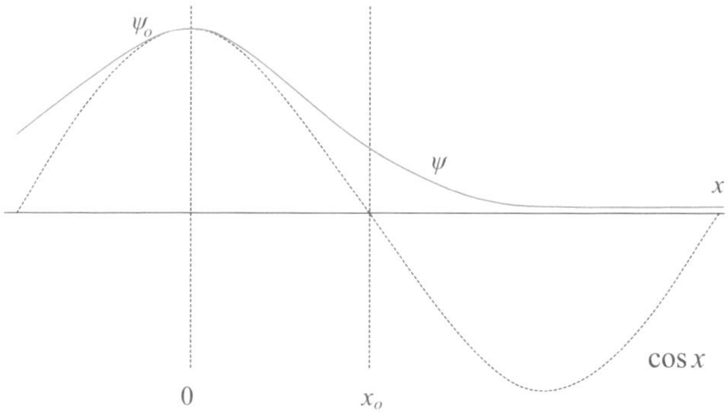

The interference energy is $E_{interference\ wave} = \int_{-\infty}^{+\infty} \psi^2(x)\, dx = \psi_o^2 \frac{x_o}{2}$.

With $\psi_o = \sqrt{force = ax}$ the integral becomes $E_{interference} = \frac{ax_o^2}{2}$;

The interference waves of the harmonic oscillator and those of the hydrogen atom differ with respect to the shape of the potential: they are convex or concave, but are similar in terms of the amplitude maximum at the origin as well as the fast decay beyond the radius of mean displacement. The interference waves of lowest energy are for:

Harmonic oscillator	$\psi_{HarmOsc} \sim e^{-x^2/2x_o^2}$
Hydrogen atom	$\psi_{Hydrogen} \sim e^{-r/r_{Bohr}}$

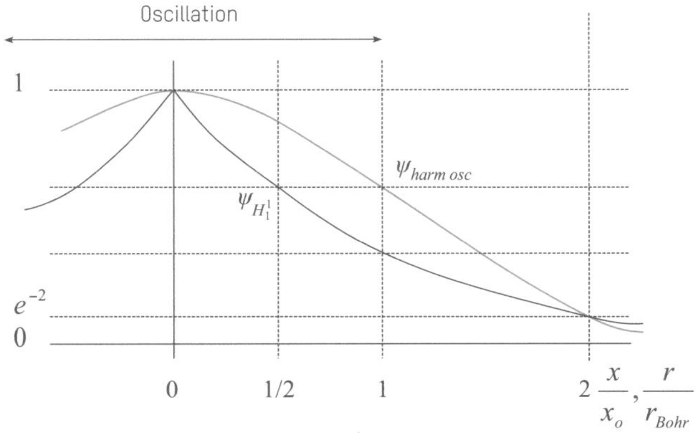

For the harmonic oscillator as well as for the hydrogen atom resonance occurs, if integral distances and angular momenta are present simultaneously:

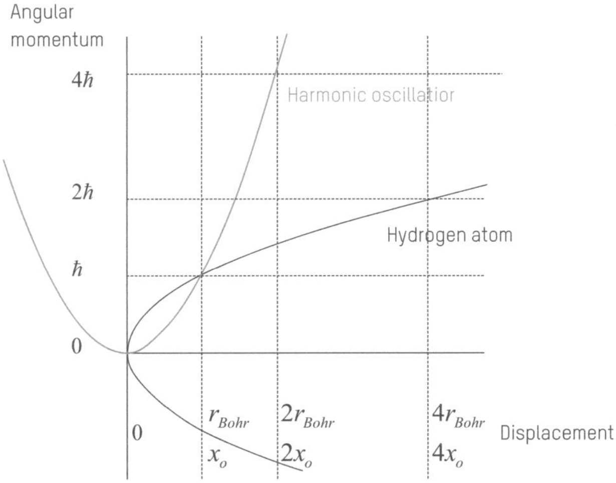

As angular momentum is intrinsically tied to energy, its invariance follows from that of the energy, and is determined in value by resonance.

9. Schrödinger Equation

The Schrödinger equation is held to be underivable in inductive physics. Deductive physics derives it in three independent ways based on

- the momentum theorem, conventionally argued: converse of *Ehrenfest's* derivation,
- the momentum theorem, freshly argued,
- the superposition of Lorentz contractions from relative velocity *and* force field.

Converse of Ehrenfest's derivation

The mean location $\bar{x}$ of a particle of mass m with interference wave ψ whose amplitude square is w is

$$\bar{x} = \int_{-\infty}^{+\infty} wxdx$$

Analogously the mean force is $\bar{F} = \int_{-\infty}^{+\infty} wFdx$.

If w moves, the corresponding flux is

$$s = \frac{\hbar}{2im}(\psi^*\psi' - \psi^{*\prime}\psi)$$

which yields $v\psi^*\psi = vw$ (v: packet velocity; in case of absent mass c).

To kinetic energy, energy is neither being added nor lost nor stored, due to which the momentum theorem as conservation principle for mean values applies a priori (in QM: Ehrenfest equation):

$$M \quad \frac{d(m\bar{v})}{dt} = \bar{F}$$

The energy of the interference wave is conserved as well. As it is proportional to w, this conservation (with s as the flux of w) conforms to

$$C \quad \frac{\partial w}{\partial t} + divs = 0$$

If C is inserted into M, as usual a wave equation forms. C and M, however, both contain ψ and ψ^*, so a little calculation cannot be avoided. First one has to write out C in ψ and group suitably:

$$C \quad \psi(\dot{\psi}^* - \frac{\hbar}{2im}\psi^{*''}) + \psi^*(\dot{\psi} + \frac{\hbar}{2im}\psi'') = 0$$

Denoting the parenthesis expressions with A and A^*, C becomes $\psi A^* + \psi^* A = 0$, with the solution

$$A = y(x)\cdot\psi$$
$$A^* = -y(x)\cdot\psi^*$$

where $y(x)$ is a yet unknown function whose determination requires M. Firstly the parenthesis expressions from C have to be inserted into A and A^*

$$A^* \quad \dot{\psi}^* - \frac{\hbar}{2im}\psi^{*''} = -y(x)\psi^*$$

$$A \quad \dot{\psi} + \frac{\hbar}{2im}\psi'' = y(x)\psi$$

to be processed further in the step after the next.

For forming $\frac{\partial(m\bar{v})}{\partial t} + \bar{F} = 0$ $\bar{v}$ is required

$$\bar{x} = \int wx dx$$

$$\dot{\bar{x}} = \bar{v} = \int \frac{\partial w}{\partial t} x dx = -\int \frac{\partial s}{\partial x} x dx = sx + \int s dx$$
$$= \frac{\hbar}{2im}\int(\psi^*\psi' - \psi^{*'}\psi)\, dx = \frac{\hbar}{im}\int \psi^*\psi' dx$$

Inserting for $\dot{\psi}^*$, $\dot{\psi}$ from A, A^* into the desired derivative

$$\frac{d(m\bar{v})}{dt} = \frac{\hbar}{i}\int(\dot{\psi}^*\psi' - \psi^*\dot{\psi}')\, dx$$

results in

$$C \to M \qquad \frac{\hbar}{i}\int \psi^*\psi \frac{\partial y}{\partial x} dx = \int F\psi^*\psi dx$$

after vanishing of terms, restructuring, partially integrating and setting equal, which means that $y(x) = \frac{i}{\hbar}\int F dx = U(x)$.

Inserting this into A or A^* yields

$$S \quad i\hbar\frac{\partial\psi}{\partial t} = -\frac{\hbar^2}{2m}\frac{\partial^2\psi}{\partial x^2} + \psi U(x)$$

$$S^* \quad i\hbar\frac{\partial\psi^*}{\partial t} = +\frac{\hbar^2}{2m}\frac{\partial^2\psi^*}{\partial x^2} - \psi^* U(x)$$

i.e. the Schrödinger equation and its complex conjugate.

In several dimensions on top of ψ'' and analogously to $\psi^{*''}$ come

– cylindrically: $+\frac{\hbar}{2im}\frac{\psi'}{r}$,

– spherically symmetric: $+\frac{\hbar}{2im}\frac{2\psi'}{r}$.

Based on the momentum theorem the Schrödinger equation is established in just five steps:

- Integrating Newton $mv = mv_o + \int_o F(x)dt$.
- Satisfying two wave mechanical conditions by substitution of
  - v with the yet unknown velocity of the wave packet, $v = \partial\omega/\partial k$,
  - v_o with the given velocity of the mass, $v_o = \hbar k/m$,
  - dt via the relation

$$dx = v_o dt \to \frac{\partial(dx)}{\partial k} = \frac{\partial v_o}{\partial k} dt = \frac{\hbar}{m} dt$$

– Inserting:

$$\frac{\partial\omega}{\partial k}=\frac{\hbar k}{m}+\frac{1}{m}\int_o F(x)\frac{m}{\hbar}\frac{\partial(dx)}{\partial k}$$

– Partially integrating over k and expanding by $\hbar$

$$\hbar\omega=\frac{\hbar^2k^2}{2m}+\int_o F(x)\,dx$$

– For ω and k from derivatives of $\psi = e^{-i(kx-\omega t)}$ inserting for the interference wave, and with $U(x) = \int_o F(x)dx$

$$i\hbar\dot{\psi}=-\frac{\hbar^2\psi''}{2m}+U(x)\psi$$

If the equation from partial integration, which is quadratic in $\hbar$, is solved for $\hbar$ (with $k = 1/\lambda\!\!\!^{-}$), then:

$$\hbar=m\omega\lambda\!\!\!^{-2}\left(1\pm\sqrt{1-\frac{2U(\lambda\!\!\!^{-})}{m\omega^2\lambda\!\!\!^{-2}}}\right)$$

Expanding this equation by $\omega/2$ the angular momentum before the parenthesis turns into the same kinetic energy E as that within the denominator inside the parentheses:

$$\frac{\hbar\omega}{2}=E\left(1\pm\sqrt{1-\frac{U}{E}}\right)\rightarrow E=\frac{(\hbar\omega/2)^2}{\hbar\omega-U}$$

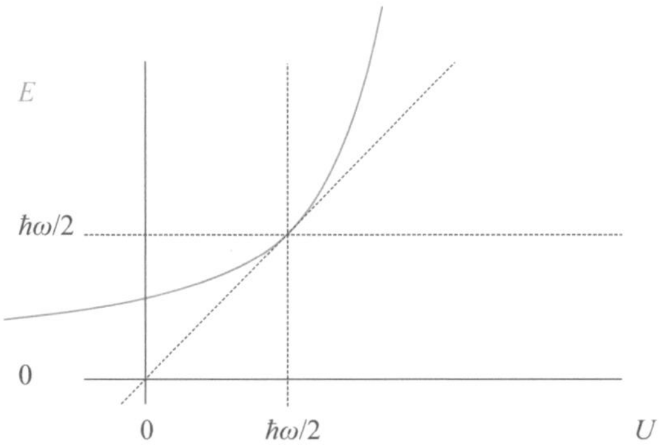

If E and U are equal, the evoked frequencies are equal, and resonance occurs: $E = U = \hbar\omega/2$. It also follows $\lambda\!\!\!^{-2} = \hbar/m\omega$. Which means: the Schrödinger equation is the momentum theorem in the form bringing to bear the interference of radiations, on which all forces are based ("eigenvalues"). Furthermore it can be inferred that E cannot be zero (save formally for $U \rightarrow -\infty$), but corresponds to the kinetic energy of a zero-point motion.

Derivation from superposition of Lorentz contractions

Closest to deductive physics is the derivation which only presupposes that the wave be understood as interference wave and mass be assigned a rest frequency $\omega_{rest} = 2mc^2/\hbar$: then the frequency ω of the interference wave of a mass with momentum $\hbar k$ within a potential $U(x)$ is determined by the following formula

$$\omega = \frac{\omega_R' - \omega_R}{2} = \frac{\omega_R}{2}\left(\frac{1}{\sqrt{1 - v^2/c^2 - 2U(x)/mc^2}} - 1\right)$$

which turns into the non-relativistic one on linearising root and fraction and yields the Schrödinger equation:

$$\hbar\omega = mc^2\left(\frac{1}{\sqrt{1 - \hbar^2 k^2/m^2c^2 - 2U(x)/mc^2}} - 1\right)$$

$$\hbar\omega \approx \frac{\hbar^2 k^2}{2m} + U(x)$$

$$i\hbar\dot{\psi} = -\frac{\hbar^2}{2m}\psi'' + U(x)\psi$$

10. Zero-point Motion Generalised

The field energy annihilated by interference is

- the new conservation quantity that QM implicitly introduces into physics;
- the cause for the energy of the hydrogen atom being only half the energy that would be given by displacement to the Bohr radius,
- the source for zero-point motion.

When a moving mass meets a force, both interference waves from relative velocity and potential field tune into a common frequency. This also happens if the mass is placed onto the zero-point of the force field. Interference energy has a maximum there that pushes the mass from the position. High velocity during transit of the zero-point contracts the wave length; the resulting displacement contraction reduces the potential energy, whereby velocity reduces again. Thus the system tunes into a sharply defined energy—the process has nothing to do with uncertainty:

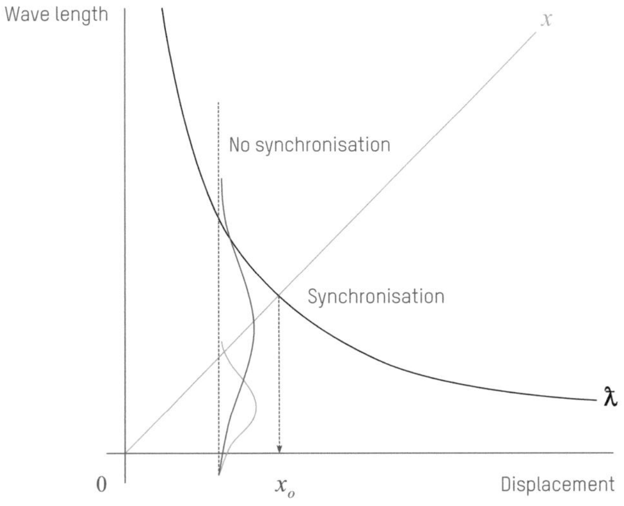

If the intersection point of the curves for x and λ is beyond x_o, the interference waves due to kinetics and due to force field do not coincide, their interference energy is larger than the minimum whereby m gets moved around until it is reached. The (negative) interference energy and the (positive) kinetic energy add up to zero: radiation energy has been transformed into kinetic energy. (cf. section on the harmonic oscillator, p. 72ff.). A force free mass at rest does not have any zero-point energy: if the spring constant in the harmonic oscillator is $a = 0$, then so is the zero-point energy.

III – Elementary Particles

The dynamics generating elementary particles in deductive physics is supported by the same continuum as the expansion of the universe. It seamlessly transitions into inertial, gravitational and electrical fields. In the chart it is symbolised by three dots:

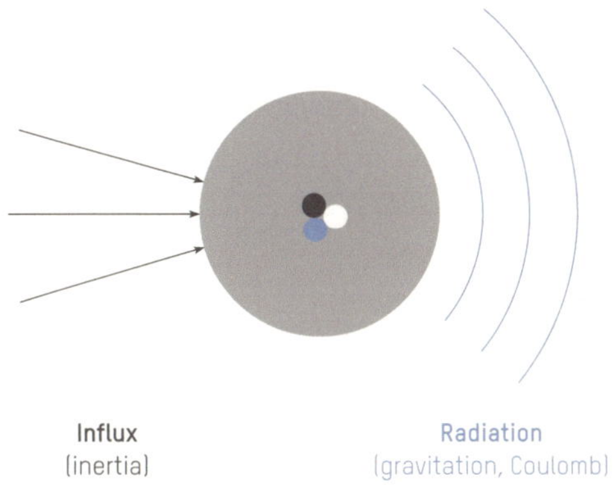

Elementary particle dynamics builds from simple to complex:

– quintessential building blocks are vortices—not unthinkable "particles" (corresponding to quarks and leptons in the standard model of inductive physics);
– these vortices cannot exist in isolation and combine to orthogonal vortex structures (the three spatial axes correspond to the three colours in the Standard Model of inductive physics);

- cohesion of the vortices is afforded by the interference annihilating a maximum of field energy at certain distances, deviation from these locations evokes restoring forces (corresponds to the strong force in the Standard Model of inductive physics);

- new conservation quantities (in addition to those from mechanics and QM) are
  - primarily: vortices
  - secondarily: vortex structures
    (correspond to quantum numbers such as iso-spin or parity in the Standard Model of inductive physics);

- the electrical field is generated by vortex structures (the corresponding cause is labelled “charge” in inductive physics).

With this structure deductive physics is in contrast to inductive physics whose building blocks—quarks and leptons—a priori contain mass, spin $\hbar/2$, charge and all their conservation quantities.

1. Vortices

In vortices the third degree of freedom of a continuum—next to flux and radiation—manifests itself: rotation. The fourth degree of freedom, rotating radiation, is exploited by the electrical field.

Determination of vortex radius and angular momentum: Vortices are fed by a flux, and the inflow is radiated. The volumes flowing in possess an angular momentum density l:

$$l = \rho(r) \cdot \omega r^2$$

(ρ *density of continuum, r radius,* ω *angular velocity*)

ρ depends on the velocity of the continuum: $\rho = \rho_\infty e^{-\omega^2 r^2/2c^2}$, by which

$$l = \rho_\infty e^{-\omega^2 r^2/2c^2} \omega r \cdot r$$

About an expected, yet unknown, innermost radius $dl/dr = 0$:

$$dl/dr = -\frac{2lr\omega^2}{2c^2} + \frac{2l}{r} + \frac{l}{\omega}\frac{d\omega}{dr} = 0$$

there $\rho = const$, by which $\frac{d\omega}{dr} = -\frac{\omega}{r}$, and the equation is satisfied for

$\omega r = c$. Synchronisation of rotation frequency ω_{mech} and quantum mechanical frequency ω_{QM} yields the innermost radius, i.e. the Compton length:

$$\omega_{QM} = \frac{E}{\hbar} = \frac{c}{r} = \omega_{mech}$$

$$\frac{mc^2}{\hbar} = \frac{c}{r} \rightarrow r = \lambda\!\!\!^{-}$$

From this follows the angular momentum

$$l = m\omega\lambda\!\!\!^{-}{}^2 = m\frac{mc^2}{\hbar}\frac{\hbar^2}{m^2c^2} = \hbar$$

This value is indeed in contrast to the spin $\hbar/2$ the Standard Model of inductive physics assigns to quarks; since it does this without direct proof. Quarks cannot be isolated, because vortices cannot exist in isolation (Helmholz), but only in order for the spin $\hbar/2$ accounting of the baryon built of three quarks, this contradiction can be passed over.

The equation for the tangential velocity ωr does not yield real values for $r/\lambda\!\!\!^{-} < 1$; at the minimum 1 is $\rho/\rho_\infty = e^{-1/2}$. Values for $\omega r/c > 1$ do not possess any physical correspondence:

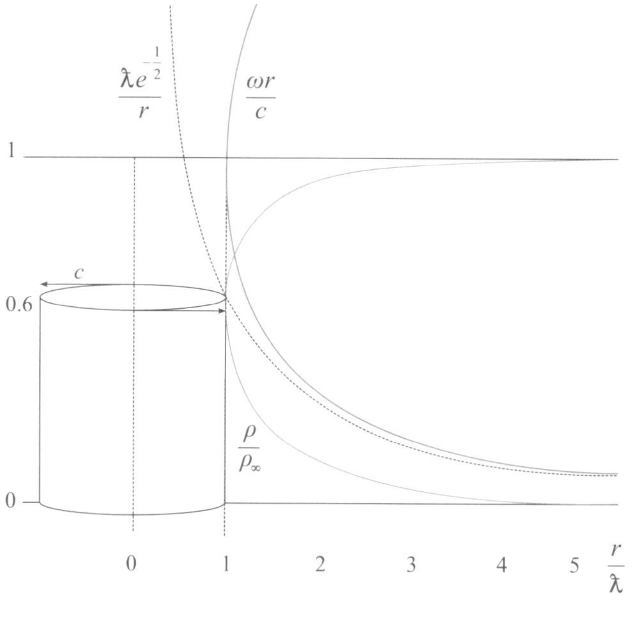

The situation is formally similar to radial influx. However, instead of the Schwarzschild radius r_S the Compton length occurs as the inner-most radius—for the proton a difference of 38 orders of magnitude:

$$\frac{\lambda\!\!\!^{-}_{proton}}{r_{Schwarzschild}} = \frac{\hbar}{m_{proton}\,c}\,\frac{c^2}{2Gm_{proton}} = \frac{m^2_{Planck}}{m^2_{proton}} = 0.85 \cdot 10^{38}$$

Black hole with radial influx:

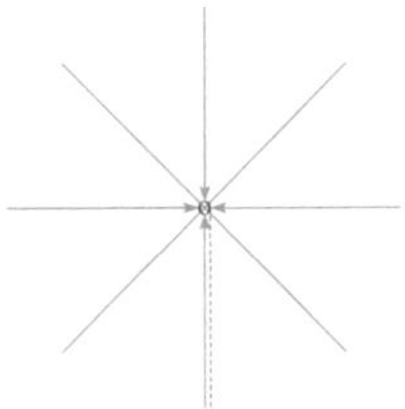

$$r_{Schwarzschild} = \frac{2Gm_{proton}}{c^2}$$

The *excentrical* influx flows tangentially to the radius ƛ, from which the radiation builds up. It is superimposed over the influx—as in the concentric case.

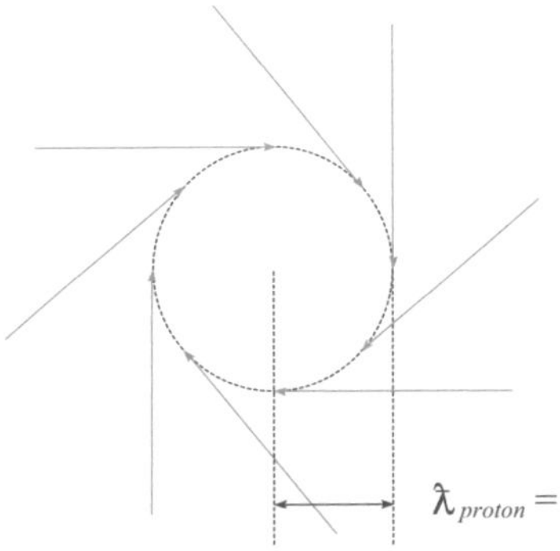

$$\lambda\!\!\!^{-}_{proton} = \frac{\hbar}{m_{proton}\,c}$$

The vortex frequency matches the one in Dirac's matrix equation, but not the rest frequency of material particles $\omega_{rest} = 2mc^2/\hbar$, which was obtained from

$$\omega_{interference} = \frac{\omega_{rest}}{2}\left(\frac{1}{\sqrt{1-v^2/c^2}}-1\right) \approx \frac{mv^2}{2\hbar} \rightarrow \omega_{rest} = \frac{2mc^2}{\hbar}$$

Thus it has to be differentiated between vortices and material particles:

		Spin	*Rest frequency*	*Radius*
–	vortex	$\hbar$	$\omega_{rest} = \frac{mc^2}{\hbar}$	$\lambda\!\!\!^{-} = \frac{\hbar}{mc}$
–	material particle	$\hbar/2$	$\omega_{rest} = \frac{2mc^2}{\hbar}$	$\lambda\!\!\!^{-} = \frac{\hbar}{2mc}$

2. Structures Composed of Vortices

Since vortices cannot exist in isolation, their dynamics have to close, e.g. by several mutually feeding each other. Their rotation generates mutual forces: for parallel axes vortices rotating in opposite directions attract each other, ones rotating in the same direction repel each other. The force is inversely proportional to the distance of the axes. With vorticity, as defined in gas dynamics, $\Gamma = 2\pi \cdot ƛ \cdot c$ as well as linear velocity reduction in the plane, $u = cƛ/r$ the force F is

$$F = \rho \cdot u \cdot \Gamma$$

$$F = 2\frac{ƛ^2 \pi \rho \cdot c^2}{r}$$

For deviation from parallelism by an angle β only the projection of the spin vectors onto each other is acting: $F \sim \cos \beta$. If the axes are mutually orthogonal, force and corresponding energy vanish, which leads to vortices aligning mutually orthogonally, if pushed together by the strong force:

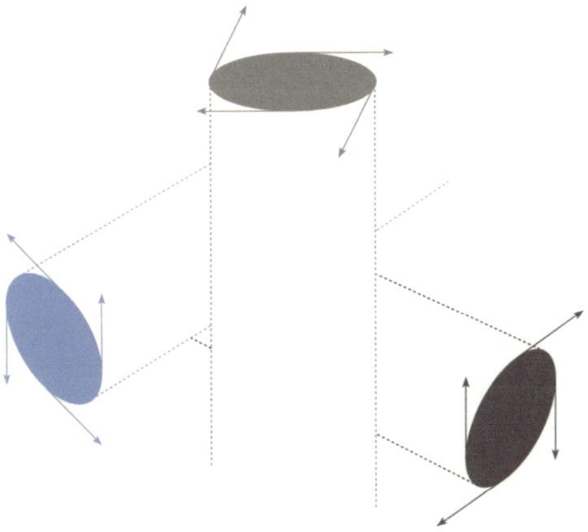

Because this tri-orthogonal configuration has minimal energy, it occurs inevitably. The three directions occur in the Standard Model as the three necessarily different colours of quarks being part of baryons.

Matrix Equation of Dirac

Rest frequency, spin, magneticity as well as anti-matter are already established in Dirac's original approach:

$$i\hbar\dot{\psi} = \frac{\hbar c\psi'}{i} + mc^2\psi$$

It combined the insight from TR that mass has a rest energy $E_{rest} = mc^2$ with that from QM that to each mass has to be assigned a frequency proportional to its kinetic or potential energy: $E = \hbar\omega$ from which results

- at rest ($\psi' = 0$): $\omega = mc^2/\hbar$,
- stationary ($\dot{\psi} = 0$): $\lambda\!\!\!^{-} = \hbar/mc$,
- ω and $\lambda\!\!\!^{-}$ combined with m to an angular momentum: $l = m\omega r^2 = m\frac{mc^2}{\hbar}\frac{\hbar^2}{m^2c^2} = \hbar.$

Dirac's matrix equation

$$\dot{\psi}(x) + \alpha_{xx}c\frac{\partial\psi(x)}{\partial x} + \alpha_{xy}c\frac{\partial\psi(x)}{\partial y} + \alpha_{xz}c\frac{\partial\psi(x)}{\partial z} + \beta_x\frac{mc^2}{i\hbar}\psi(x) = 0$$

obtains its declarative force from the Klein-Gordon condition to which it is subject on every axis: the coefficients must satisfy

$$-\dot{\psi}^2 = -c^2\psi'^2 + \omega_R^2\psi$$

respectively from which follows $\alpha_{ii}^2, \beta_i^2 = 1$. Since Klein-Gordon is the algebraic transformation of $E = mc^2/\sqrt{1-v^2/c^2}$ into $E^2 = m^2v^2c^2 + m^2c^4$, the energy of the object to be described increases proportionally to $1/\sqrt{1-v^2/c^2}$, by which Lorentz invariance is satisfied a priori.

From Klein-Gordon emerges via the quantisation $(\omega^2 = k^2c^2 + \omega_{rest}^2)$ the equivalence of positive and negative values for frequency and wave number, from which leaps the prognosis of the anti-particles (the energy is equal, whether a structure be turning right or left). And for small velocities v the de Broglie-Einstein relations can be derived from the quantisation:

$$(\omega - \omega_{rest})(\omega + \omega_{rest}) \approx \Delta\,\omega \cdot 2\,\omega_{rest} = k^2c^2$$

$$\frac{\Delta\omega}{k} = kc^2 \frac{\hbar}{2mc^2} = \frac{\hbar k}{2m} = \frac{v}{2} = v_{Phase}$$

from which $k = \dfrac{mv}{\hbar}$ as well as $\Delta\omega = \dfrac{mv^2}{2\hbar}$ result.

Structure Variations

For two vortices joined orthogonally there are two possible structures *d* and *u* (in the style of the Standard Model denoting different quarks in that way) that are not equivalent under rotation:

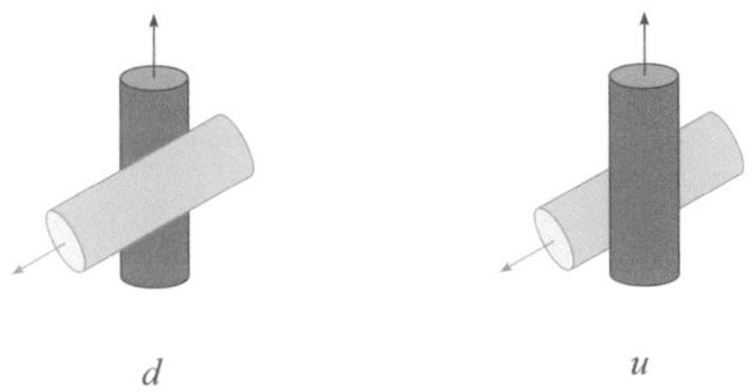

For the third vortex there are four possible relative positions

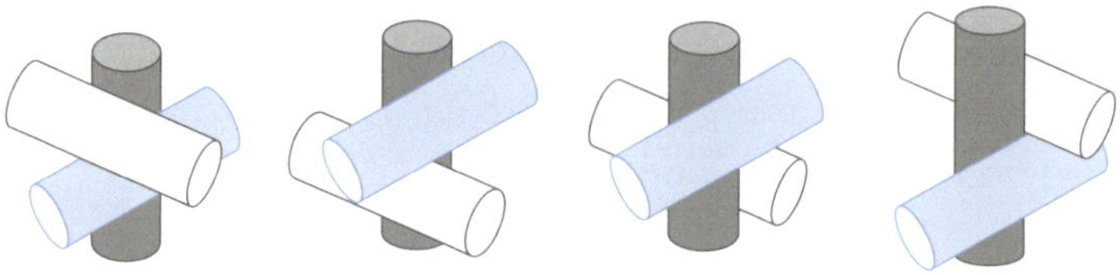

dud *ddd* *dud* *duu*

The vortex structures either feed one another or themselves:

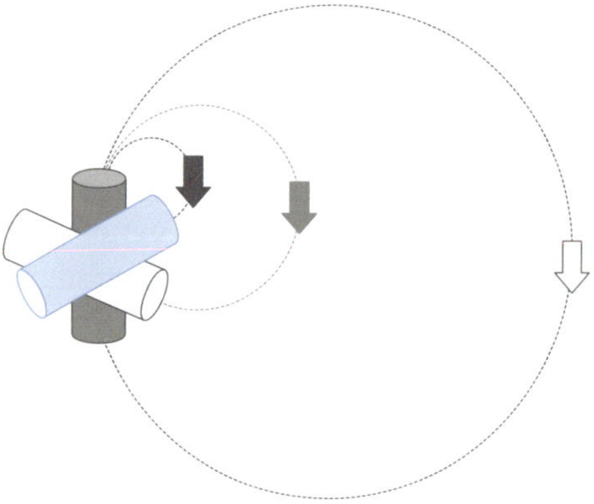

By this exchange the vortices primarily act *within* the closed structures. Externally the structure as a whole acts by

- its influx (inertia),
- the corresponding radiations (gravitation, Coulomb),
- its spin of $\hbar/2$ (double frequency in comparison to the individual vortices).

Leptons: The Standard Model of inductive physics holds leptons not to be composite, however, in deductive physics leptons have to be assigned an influx field, because they possess inertia. And as only tri-orthogonal vortex structures have influx, leptons have to be assigned three vortices, which processes like the β-decay confirm. Thus leptons do not differ in structure from baryons; ordered by increasing energy together with proton and neutron:

	Mass, MeV	*life span*
e	0.511	∞
μ	105.6	$0.2 \cdot 10^{-6}$
p	938.2	∞
n	939.5	$0.9 \cdot 10^{3}$
τ	1777	$0.3 \cdot 10^{-12}$

3. Strong Force

The strong interaction is based on the principle illustrated by interference of one wave train of *2π (blue in the chart below)* with an oncoming wave train *(grey)* of the same absolute amplitude (= 1), wave length and frequency. For a phase shift of δ the amplitude of the interference wave is

$$A = \sin x + \sin(x - \delta)$$

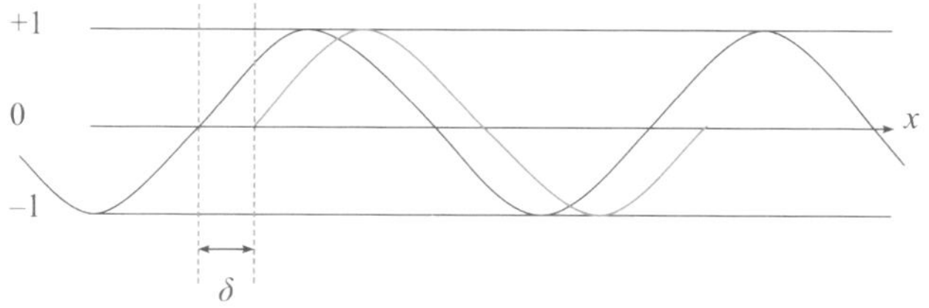

The energy of the interference wave train,

$$E \sim \int_0^{2\pi} A^2 dx = 2\pi\left(1 + \cos\delta\right)$$

as well as the force as resistance against shift

$$F \sim \frac{dE}{d\delta} = -2\pi \sin\delta$$

The particle generating the wave train abides within the phase shift where interference energy is minimal: at $\delta = \pi$. There the force F is pushing, thus it serves to lock in:

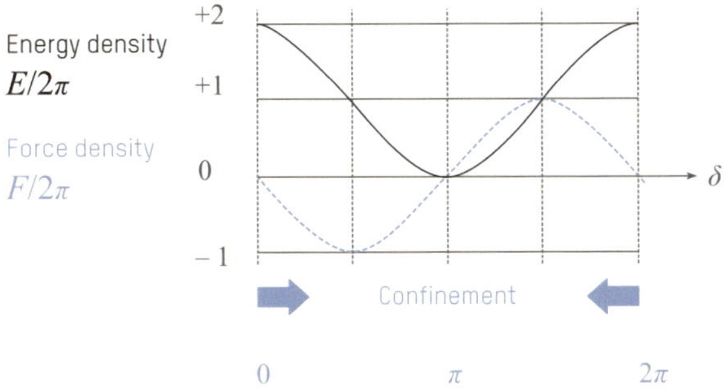

If this principle is transferred to vortices whose radiations interfere with each other at Coulomb length distance, then with the frequency $\omega_R = mc^2/\hbar$ and the wave length $\lambda = mc/h$ distances result whose magnitudes match the experimentally established *confinement*. In the case of the proton mass

$$\frac{\lambda_{proton}}{2} = 0.66 \cdot 10^{-15}$$

$$confinement \sim 0.6 \cdot 10^{-15}$$

Yukawa Potential

The exchange between the vortices reduces inverse-proportionally and exponentially with the distance between them, which the simplified case of influx resulting from suction illustrates (without considering rotation, spin axis parallel z; velocity u in direction of z, v in direction of r).

From $u^2 + v^2 = const$ follows the Euler equation M (ignoring compressibility which only becomes effective r nearing $\lambda\!\!\!^-$):

$$M \quad \frac{\partial u}{\partial z} = -\frac{v}{u}\frac{\partial v}{\partial z}$$

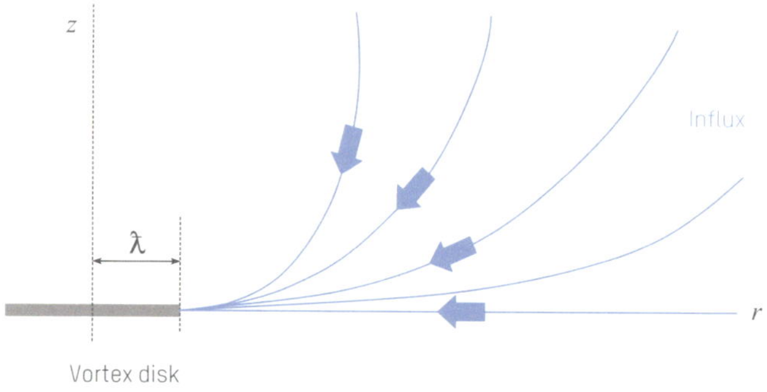

The continuity equation reads (again ignoring compressibility):

$$C \quad \frac{1}{r}\frac{\partial(vr)}{\partial r}+\frac{\partial u}{\partial z}=0$$

Inserting $M \rightarrow C$ yields

$$\frac{1}{r}\frac{\partial(vr)}{\partial r}+\frac{v}{u}\frac{\partial v}{\partial z}=0$$

With $\psi_r = v$; $\psi_z = u$ this becomes $\psi_{rr}+\frac{\psi_r}{r}+\psi_r\frac{\psi_{rz}}{\psi_z}=0$.

For large z results $\psi_{rz}/\psi_z=1/z$ leading to $\nabla^2\psi_{sphere}=0$. For small r the influx towards an individual vortex quickly increases with decreasing r because of the influx from below and above the vortex disk. Hence the potential as well: with the lateral influx (mathematically: "source term") set proportional to the potential itself with factor k^2 the radial component is determined by $\Delta\psi - k^2\psi = 0$. Here k^2 is initially an arbitrary constant. The spherical solution is:

$$\psi=\frac{e^{\mp(1-kr)}}{kr}\psi_o$$

and corresponds to the Yukawa potential. If the influx velocity for $kr = 1$ is set to c, $\psi_o=c/2k$ turns out.

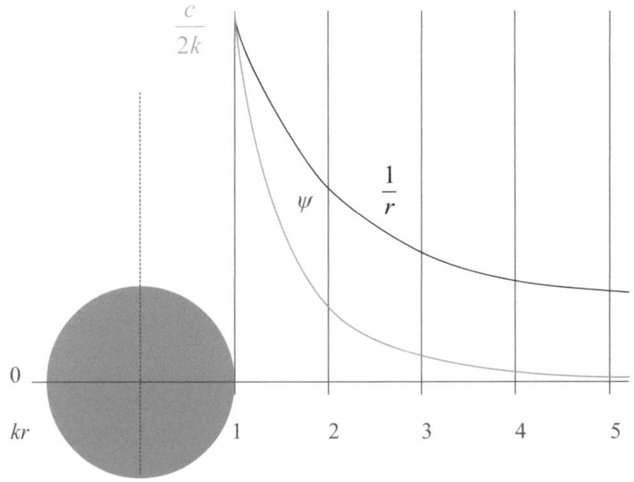

The Yukawa approximation primarily concerns the influx and secondarily the radiation superimposed over it. It explains why the strong forces (order of magnitude mc^2/r at the origin) between the vortices rapidly declines with distance. After 89 radii the potential declines to the order of magnitude of the gravitational potential (factor $e^{1-89}/89 = 0.2 \cdot 10^{-38}$); if radiation were to decline exactly according to Yukawa, it would turn into gravitational radiation at a radius r_G.

At r_G radiation reaches equilibrium with the influx towards the entire structure. It is being fed continuum by it, no longer by the vortices within the structure. Setting Yukawa and gravitational potential equal for a proton with mass m_p yields the mentioned factor:

$$\frac{\hbar c e^{1-r_G/\lambda\!\!\!^{-}_p}}{r_G} = \frac{G m_p^2}{r_G} \rightarrow \frac{r_G}{\lambda\!\!\!^{-}_p} = 89.025$$

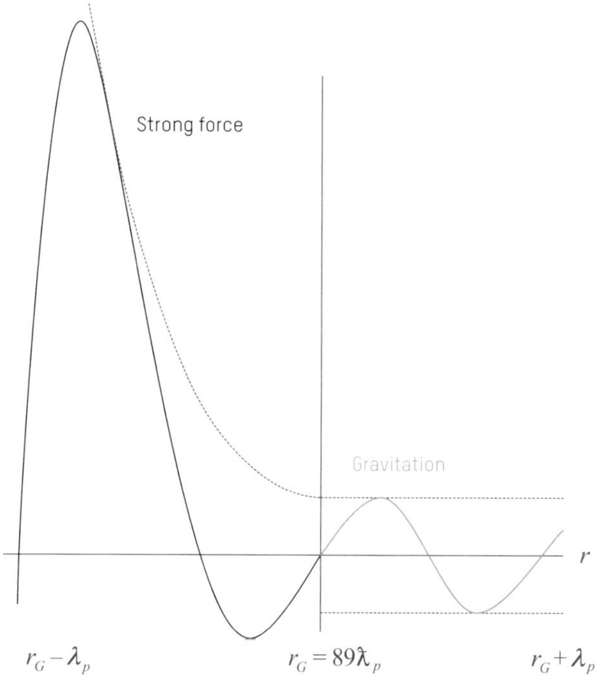
Strong force
Gravitation
r
$r_G - \lambda_p$
$r_G = 89\bar{\lambda}_p$
$r_G + \lambda_p$

4. Energy Levels of Vortex Structures

Three factors determine the various energy levels of tri-orthogonal structures

a) Angular momenta of the three vortices,
b) Excitation of the entire structure,
c) Angular momentum of the entire structure.

a) Contribution of the angular momentum of the vortices

On each axis there are angular momenta $n\hbar$ with $n \geq 1$ for durations of order 10^{-24} s. The energy difference of 252 MeV between p and Σ is the contribution of the additional angular momentum on *one* axis, the difference of 379 MeV between p und Ξ on *two* axes (on average the energy increase by one $\hbar$ corresponds to almost two-thirds of the base energy).

The Standard Model of inductive physics fails to explain the energy levels of the short-lived particles, which it finds in experiment, and employs a quantum number for this, S:

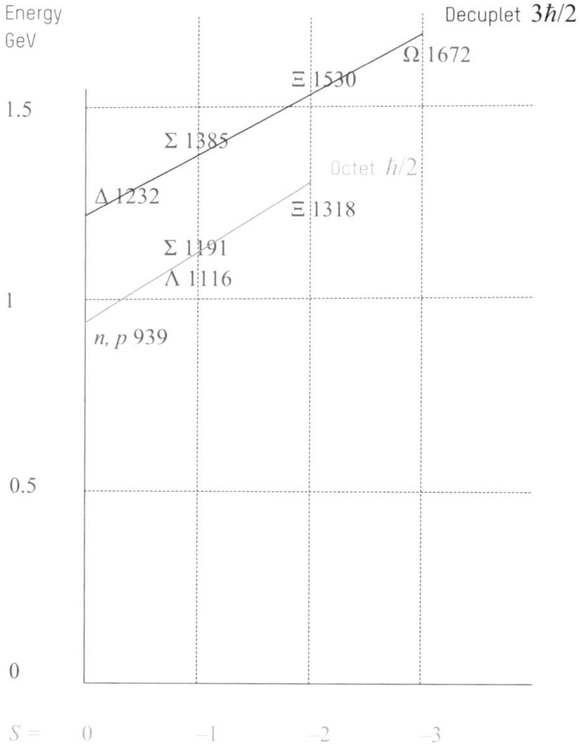

b) Contribution of the excitation of the entire structure to the energy level

The radiation energy annihilated by interference, $E_{interference}$, is at its greatest for a phase shift of half a wave length. Consequently this is the mutual distance the vortices occupy unperturbed. In contact, distance zero, interference energy has the positive maximal value, which is why contact does not occur.

The amplitude of the interference wave is inversely proportional to the vortex distance ("Yukawa" does not act on waves). Thus interference energy relates to rest energy as follows:

$$\left|E_{Interference}\right| \approx \frac{\hbar c}{\lambda\!\!\!^{-}(n+1/2)} \qquad d, \textit{number of wave lengths}$$

The measured values for the excited states of the proton correspond approximately to those from this formula:

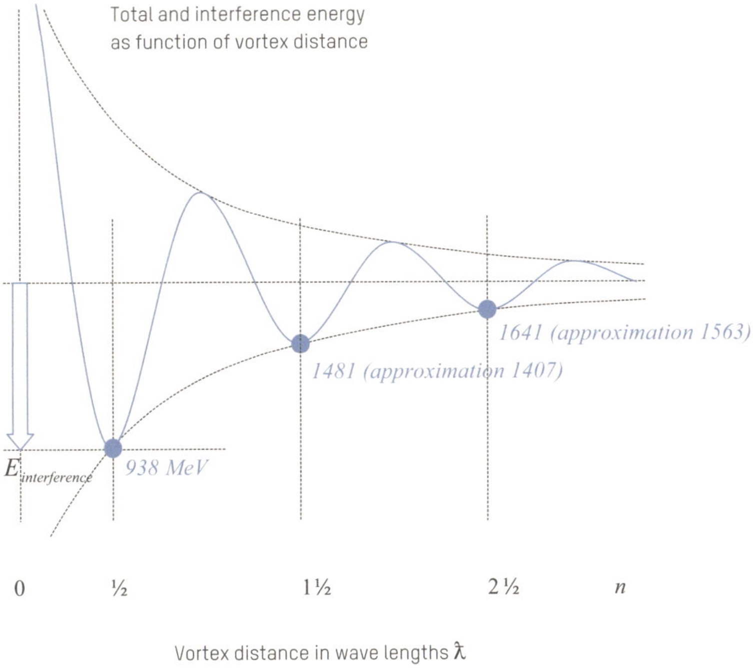

c) Contribution of the angular momentum of the entire structure to the energy level

The vortex axes in ground state have distances of two ƛ plus ƛ/2 from maximising interference energy. The distance r of the three vortex axes from the common pivot point is $r = 5ƛ/\sqrt{8}$. If the structure rotates, it has an additional angular momentum of $l = m\omega r^2$ with value $n\hbar$ on top of the spin

$$m\omega_{mech}\left(\frac{5ƛ}{\sqrt{8}}\right)^2 = n\hbar$$

The additional energy gained via angular momentum is

(with $ƛ = \frac{\hbar}{mc}$)

$$E_{rot} = \hbar\omega_{mech} = \hbar\frac{n\hbar}{m}\frac{8}{25ƛ^2}$$

$$= \frac{n\hbar^2 m^2 c^2}{m\hbar^2} = \frac{8n}{25}mc^2$$

For the elementary particle Δ the total energy is ($n = 1 \rightarrow l = 3\hbar/2$ as well as $mc^2 = 938\ MeV$)

$$E = 938\left(1 + \frac{8}{25}\right) = 1238\ MeV$$

Measurements yield 1232 *MeV.*

The real situation is more complex than illustrated in these approximations, as all three vortices forming the proton act on each other, apart from effects of "Yukawa" and electromagnetic potential.

5. Conservation Quantities

Conservation quantities upon which all physics is based are in

- *mechanics:* energy, linear/angular momentum, charge;
- *quantum mechanics:* spin, interference energy (in inductive physics only implicitly, however, it is treated as the new invariant),
- *elementary particle physics:* vortices and structures (corresponding to quarks, leptons and quantum numbers in the Standard Model).

The *vortex* is the primary conservation quantity in deductive physics (its conservation is founded on conservation of quantity and potential in resonance). As vortices are only viable in structures of several vortices, the secondary conservation is that of vortex *structures.*

The *beta decay* illustrates the principle of structure conservation: the neutron turns into a proton by dislocating one vortex of the neutron, in the chart: by

- removing the dark grey vortex and inserting a new vortex (light grey)

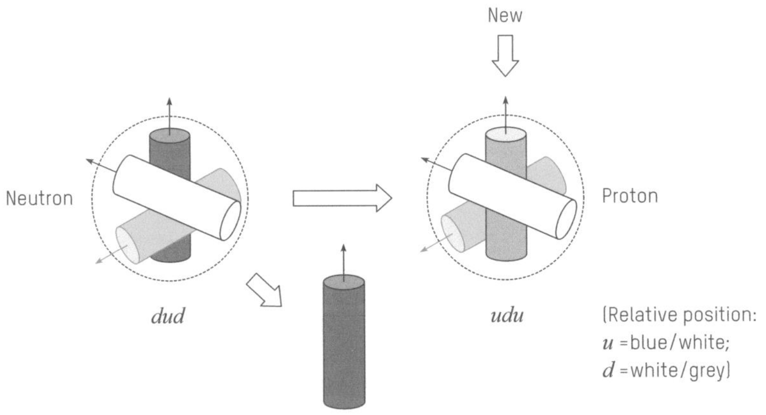

– supplementing the removed vortex with two new ones (light grey): the electron ensues

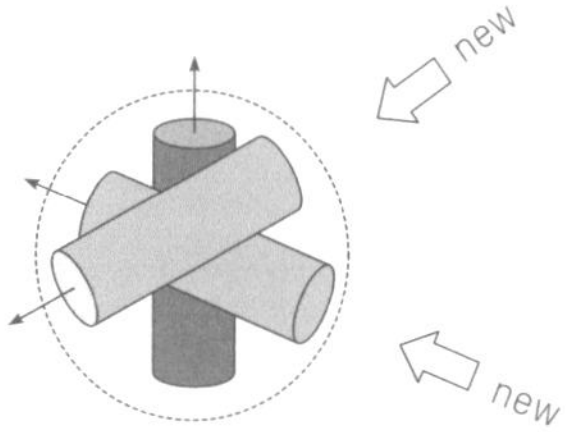

– compensating the one new vortex of the proton (grey) as well as the two new ones of the electron (grey) by three new counter oriented ones (white): the electron-anti-neutrino ensues

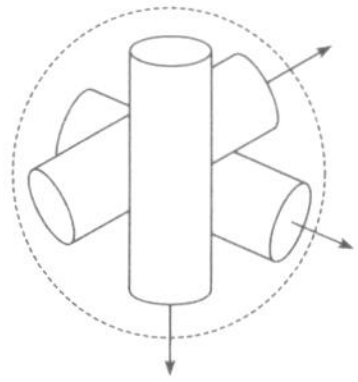

Neutrinos are

– "anti-structures" of "anti-vortices",
– built from tri-orthogonal vortices like baryons and charged leptons with spin $\hbar/2$,
– not substitutable in decay processes with members of the alternative neutrino family,
– massless: they move away from the process with *c* and therefore do not build up influx (as the photon consisting of one vortex travelling at the speed of light),
– without charge (for the same reason).

These structure variants of deductive physics correspond to Standard Model combinations of two quark types, *d* and *u*, three of which are required for a baryon:

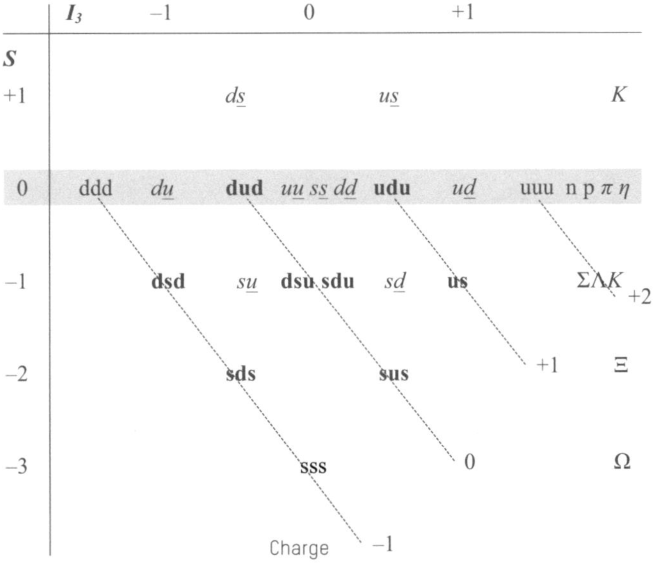

The mystery of the "vanished anti-particles" resolves as follows: In big bang vortices form first. Their angular momentum adds to zero, since with each vortex a counter rotating "anti-vortex" forms simultaneously.

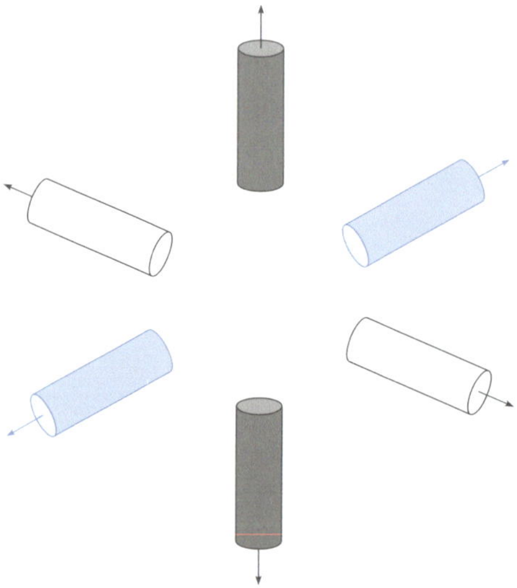

However, these do not fuse to anti-particles, but to neutrons:

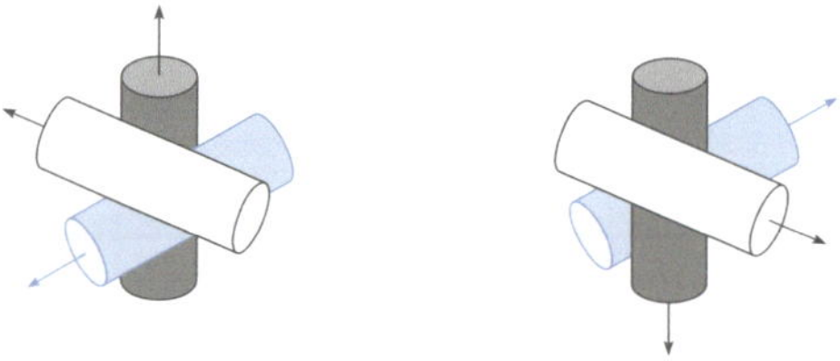

dud *dud*

The abundance of particles compared to anti-particles, specifically electrons compared to positrons, in the universe is determined by the parameters of the beta decay.

6. Comparison of Deductive and Inductive Elementary Particle Physics

DEDUCTIVE PHYSICS	INDUCTIVE PHYSICS
Building Blocks	
Vortex with spin $\hbar$	Quarks with spin ½ Leptons with spin ½
Two distinct topological relations	Two distinct quark families *u, c, t* and *d, s, b*
Structure	
Vortices are pulled together and aligned orthogonally on three axes (maximal annihilation of field energy) by the strong force	Baryons require three quarks of different colour
Spin $\hbar/2$ of the entire structure at double rotation frequency compared to individual vortex	Spin $\hbar/2$ as sum of spins of vortices
Leptons are composed of three vortices of spin $\hbar/2$	Leptons are fundamental particles, not composite, with spin $\hbar/2$
Inverse rotation orientation of vortices	Anti-particles

DEDUCTIVE PHYSICS	INDUCTIVE PHYSICS
Short-lived counter rotating vortices annihilating each other before completing full rotation about their axis	Meson

<u>Strong Force</u>

Interference annihilates field energy and exiting from locations of maximal energy annihilation evokes restoring forces	Gluon exchange

<u>Electroweak Force</u>

Multistep, thus delayed, modification of vortex structures	Boson exchange
No genuine force	Interpretation of decreasing probabilities as forces
Interference wave with annihilated field energy, thus integral spin; analogous to H_1^1 atom:	Boson in general, gluons in particular

$$E_{total} = E_{pot} + E_{interference}$$
$$= - e^2/r_{Bohr} + e^2/2r_{Bohr}$$
$$\omega_{Bohr} r^2_{Bohr} m_e = \hbar$$

DEDUCTIVE PHYSICS	INDUCTIVE PHYSICS
Energy Levels	
Variation of angular momenta on each spatial axis	Genuine particles for every energy level with distinct quantum number S
Conservation Quantities	
Primary: Conservation of sum of rotation about each of the three spatial axes	Colours, spin, angular momentum
Secondary: Structure topology	Baryon/lepton number, quantum numbers C, P, T, S and combinations of these
Structure modification has to be compensated topologically	Quark families cannot be inserted cross-wise, substitution only per $u \rightarrow c \rightarrow t$ $d \rightarrow s \rightarrow b$
Structure generates the electric field ("charge" does not have inertia, as only its electric field is real, and that is a radiation)	Charge

DEDUCTIVE PHYSICS	INDUCTIVE PHYSICS
Unification of Forces	
Forces are of an entirely different nature and unification is not necessary	Analogues to "exchange of virtual photons" for interpreting the Coulomb force: bosons, gluons, gravitons
Higgs Field and Particle	
Influx field of mass dynamics	Higgs field
Does not need any source—it is constitutive part of mass dynamics	Its source is a separate particle, the Higgs boson

Addenda to this as follows: From the Dirac equation ensues the Lagrange energy density of the field

$$L = \partial_\mu \Phi^+ \partial^\mu \Phi - \mathrm{m}^2 \Phi^+ \Phi$$

which transformed and expressed in spherical terms reads

$$\omega^2 \Phi + \nabla^2 \Phi - \omega^2_{\mathrm{rest}} \Phi = 0$$

At rest $\omega = \omega_{rest}$, from which follows:

$$\nabla^2\Phi = 0$$

$$\Phi = \frac{const}{r}$$

which corresponds to the influx field in the dynamics of deductive physics. There is no space for an additional particle, the Higgs particle, supposedly generating this field (and detected in 2012 at the LHC in Geneva). The Standard Model in its quantum mechanical descent only deals with radiation. When spelling out its mathematical consequences it inevitably encounters the influx feeding the radiation. It correctly interprets this as an infinite field in space—but it incorrectly seeks a separate particle as the source of this field.

IV – Electromagnetism

In deductive physics electrical fields are generated by elementary particle dynamics. Inductive physics assigns "charge" as cause to electromagnetic fields and locates it in the quark as one or two thirds charges.

1. Principle of the Electric Field

In deductive physics:

- The electrical field is a radiation. It decreases with relative velocity v in its direction with $(1-v^2/c^2)$, and increases orthogonally with $(1 - v^2/c^2)^{-1/2}$; because it is superimposed onto continuum flux like the gravitational field.

- The frequency of the electrical field of the electron is its rest frequency, which for example follows from the frequency of the interference wave of the hydrogen atom $\omega_{Bohr} = e^2/\hbar r_{Bohr}$. This originates from the following relation into which the frequency of the electrical field $\omega_{Coulomb}$ is initially inserted as unknown:

$$\omega_{Bohr} = \frac{e^2}{\hbar r_{Bohr}} = \frac{\omega_{Coulomb}}{2}\left(\frac{1}{\sqrt{1-2e^2/r_{Bohr}m_ec^2}}-1\right)$$

$$\approx \frac{\omega_{Coulomb}}{2}\left(1+\frac{e^2}{r_{Bohr}m_ec^2}-1\right)$$

$$\omega_{Coulomb} = \frac{2m_ec^2}{\hbar} = \omega_{rest}$$

– The radiation has twist, as the influx towards a vortex arrives tangentially. The nature of "rotation radiation" becomes evident in that an electrical field only interacts with an electrical field—analogous to gas dynamics, where vortex only acts on vortex:

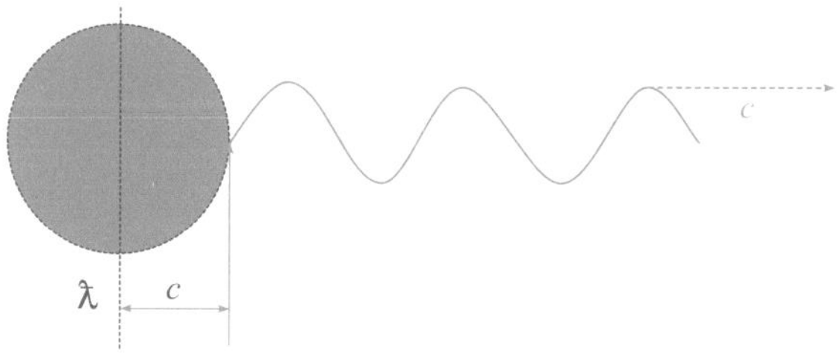

– The source of the electrical field of the electron is on a *sphere* of radius $ƛ_e$ which follows from the magnetic moment, l_m,

defined as $\quad l_m = \omega r^2 e$ (with $e = q/\sqrt{4\pi\varepsilon_o}$),

inserted $\quad \omega = \omega_{rest,e} = 2m_ec^2/\hbar$,

$$r = ƛ_e = \frac{\hbar}{2m_ec}$$

whence $\quad l_m = \frac{2m_ec^2}{\hbar}\frac{\hbar^2}{4m_e^2c^2}e = \frac{\hbar e}{2m_e}$ results.

– The emission organises itself (virtual photons with radial spin axis). The radius of the rotating radiation packets is constant, their density varies (like the density of the constituents of a gas):

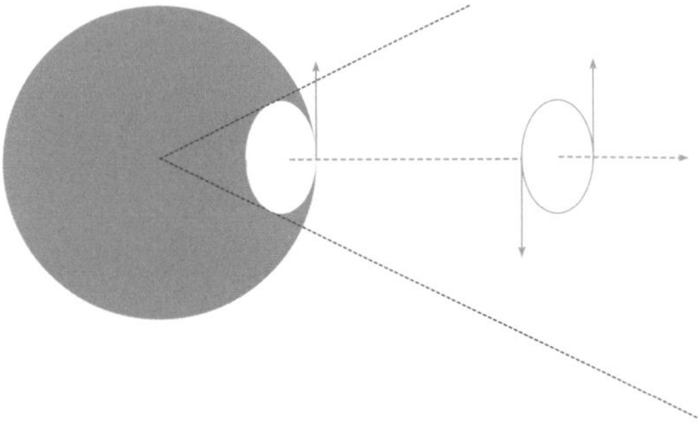

– The packets form "tubes" annihilating or pushing aside one another—depending on rotation orientation which becomes clear in the field lines:

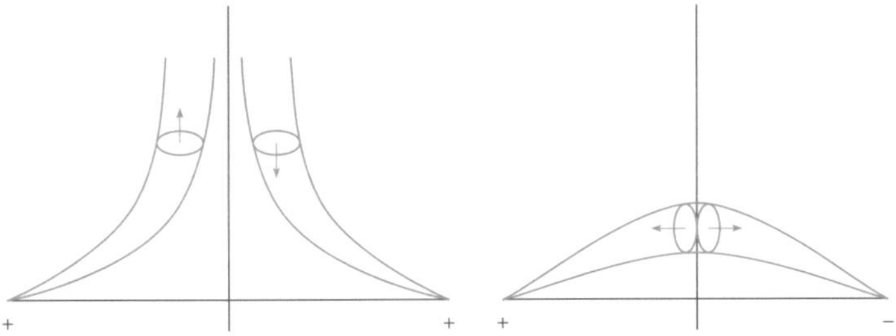

– Even inside the neutron there is electrical radiation (manifest in the magnetic moment, whose value is two-thirds of that of the proton), but it is closed. In the proton it is open (goes out) and for example in the hydrogen atom is closed in cooperation with the electron.

2. Rotation Radiation

The electrical field is formed by "virtual photons" which the mass dynamics is emitting. The virtual photon does not present a particle, but a radiation packet. Since it rotates and propagate with c, it can be called rotation radiation:

Rotation radiation

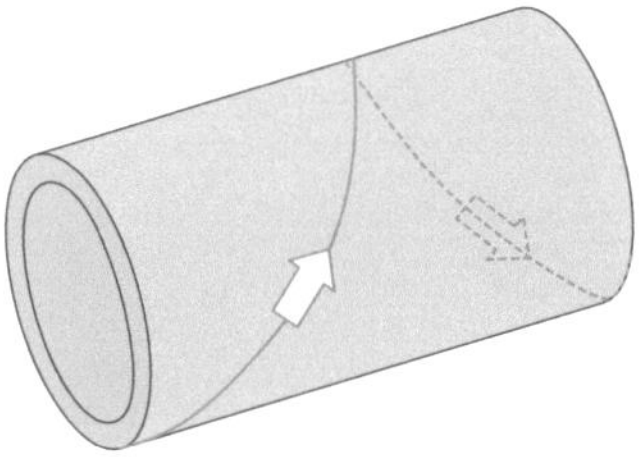

Wave equation for this rotation radiation. The radius of the packet is determined by resonance (angular momentum $\hbar$). At this given radius continuity and Euler equation, C and M, have to be determined (ω here is angle, not frequency)

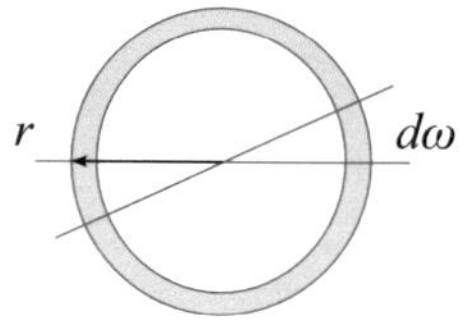

$$v = r\dot{\omega}$$
$$ds = rd\omega$$
$$dI = d\rho/\rho$$

$$C \quad \frac{\partial v}{\partial s} = -\frac{1}{\rho}\frac{\partial \rho}{\partial t} \qquad \text{from this } \frac{\partial \dot{\omega}}{\partial \omega} = -\dot{I}$$

$$M \quad \frac{\partial v}{\partial t} = -c^2 \frac{1}{\rho}\frac{\partial \rho}{\partial s} \qquad \text{from this } r\frac{\partial \dot{\omega}}{\partial t} = -\frac{c^2}{r}\frac{\partial I}{\partial \omega}$$

Differentiating once with respect to t and once with respect to ω

$$C \quad \frac{\partial^2 \dot{\omega}}{\partial \omega^2} = -\frac{\partial \dot{I}}{\partial \omega}$$

$$M \quad \frac{\partial^2 \dot{\omega}}{\partial t^2} = \frac{c^2}{r^2} \frac{\partial \dot{I}}{\partial \omega}$$

and inserting mutually yields the wanted wave equation

$$\frac{\partial^2 \dot{\omega}}{\partial t^2} = -\frac{c^2}{r^2} \frac{\partial^2 \dot{\omega}}{\partial \omega^2}$$

3. Maxwell Equations

The electrical field consists of a flux of "virtual photons", rotating wave packets with

- spin $\hbar$,
- velocity c,
- radius $ƛ$,
- frequency $\omega_{photon} = c/ƛ = mc^2/\hbar$:

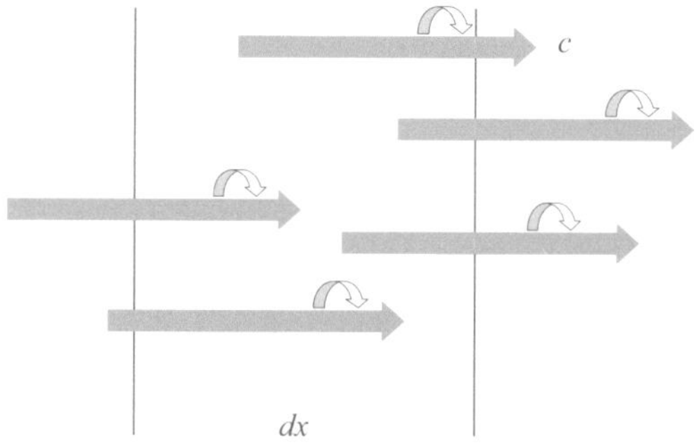

From this notion of the electric field the electromagnetic Maxwell equations are derived in a similar manner to the derivation of the field equation for the vortex free continuum: from conservation of spin density and spin flux.

Conservation of spin density: spin cannot disappear and if the photon flux of density E decreases by dE over the interval dx, then this triggers a spin dB compensating it—as for the gyroscope:

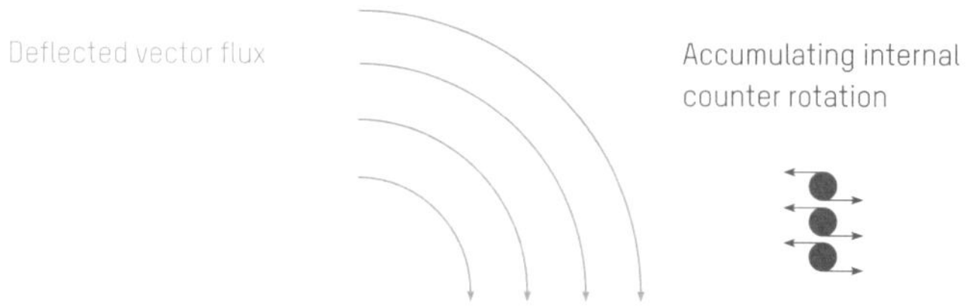

Expressed mathematically:

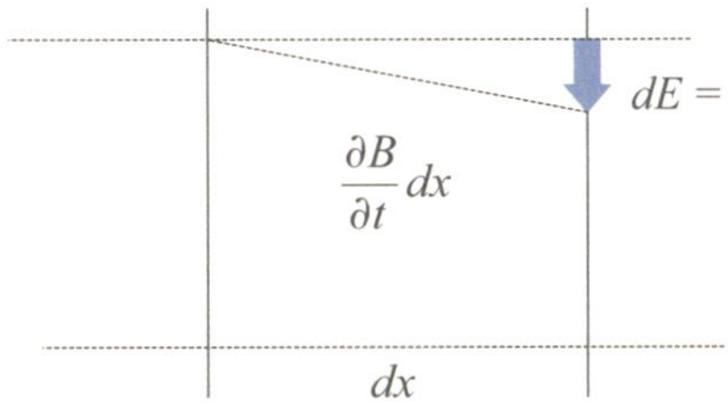

Since the substrate (which is being reduced or consumed) is spin with horizontal orientation, thus consisting of vectors, instead of *dE/dx* → *rotE* has to be inserted:

$$C \quad \frac{dE}{dx} + \frac{\partial B}{\partial t} = 0 \rightarrow rotE + \frac{\partial B}{\partial t} = 0$$

Conservation of spin flux: If the density of horizontal spin at the right hand boundary is instantaneously being reduced by *dB*, the outflux is immediately reduced by *cdB* which spread across the entire interval over time yields an average flux reduction:

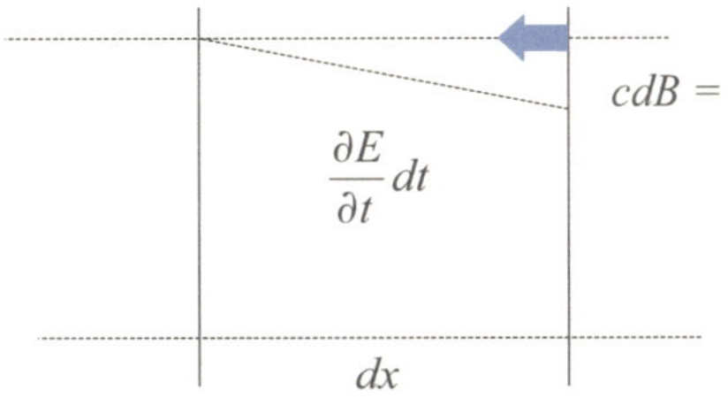

As the reduction across the entire interval is reached after *dt*, for *dt* → *dx/c* has to be inserted, again with *dB/dx* → *rotB*, with which

$$M \quad \frac{\partial E}{\partial t} - c^2 \frac{dB}{dx} = 0 \rightarrow \frac{\partial E}{\partial t} - c^2 rotB = 0$$

Thus the electromagnetic Maxwell equations have been derived from the conservation of spin density and flux:

$$C \quad rotE + \frac{\partial B}{\partial t} = 0$$

$$M \quad \frac{\partial E}{\partial t} + c^2 rotB = 0$$

The nature of B as “storage” also manifests itself in the total energy of the field:

$$E^2 + c^2B^2 = const$$

likewise the hermitian nature of the field equations: if the amplitude of an electromagnetic wave vanishes, its energy resides in the concealed (appearing imaginary, phase shifted) amplitude of the magnetic field $E = icB_{const = 0}$.

The Maxwell equations can also be reached by analogy reasoning: in the equations for the vortex free continuum

$$C \quad \frac{\partial(\rho v)}{\partial t} + c^2 \frac{\partial \rho}{\partial x} = 0$$

$$M \quad \frac{\partial(\rho v)}{\partial x} + \frac{\partial \rho}{\partial t} = 0$$

substitutions have to be made as follows

$$\rho v \rightarrow E$$
$$\rho \rightarrow B$$
$$grad \rightarrow rot$$

The continuum transmitting the perturbations, whether pressure or rotation waves, is the same:

Electromagnetism	*Gravitation*
vortex density	density of continuum
vortex flux	vortex free flux
discrete constituents ("virtual photons")	continuous density

4. Gauge Theories

Gauge theories are generalizations of the Maxwell equations. Arbitrary variable quantities are added to the potential and the momentum within the equations of motion: to the potential $V(x)$ the product $q\Phi(x,t)$, to the momentum $\vec{p}$ the product $q\vec{A}(x,t)$. Thus three additional terms appear on top of $-gradV$ in the equation of motion

$$\frac{\partial\vec{p}}{\partial t} = -gradV + qv \times rot\vec{A} - q\left(\frac{\partial\vec{A}}{\partial t} + grad\Phi\right),$$

and the "arbitrary" quantities turn out as

$$-\left(\frac{\partial\vec{A}}{\partial t} + grad\Phi\right) = E, \qquad \text{electric field,}$$

$$rot\vec{A} = B, \qquad \text{magnetic field.}$$

If the equation of motion is not to be changed by the addition of the three quantities $q, \vec{A}, \Phi$, then $\partial\vec{A}/\partial t + grad\Phi = 0$ has to hold, which renders $rot\vec{A} = 0$ as well. With $\vec{A}$ a new conservation quantity has manifested itself: the vector potential. In deductive physics it results from the following: the vortex density $\vec{\rho}$ corresponds to the magnetic field B and the vortex flux $\vec{\rho}v$ to the electric field. With this the vector potential becomes $\vec{A} = \int Edt = \int\vec{\rho}vdt = \int\vec{\rho}dx$, the analogue of the potential $\Phi = \int\vec{v}dx$. $\partial\vec{A}/\partial x$ again gives $\vec{\rho}$.

That introducing "arbitrary" variable quantities should lead to these results is not surprising: in deductive physics there is only a single degree of freedom for adding to the scalar potential $V(x)$ one which is not again a scalar potential: rotation—and the electromagnetic fields are rotational fields, as shown on p. 118. Generalized: Gauge theories introduce rotation into the field equations, specifically with the potential mc^2 in the Dirac matrix determining the rest frequency $\omega = mc^2/\hbar$ of the vortices.

Fundamental treatment recognizes only imbalances of density and velocity as degrees of freedom of the continuum; density impacts velocity and vice versa. Rotation is already—like influx and radiation of the mass dynamic—a spatial organisation thereof; rotation impacts rotation. Based on this are the further abstractions force, momentum, energy, spin, charge, etc. Rotation is only in resonance stable, which leads to angular momenta $\hbar$ and in the SM to the quantisation of the fields. Stable rotation itself is based on the dynamic equilibrium of pressure and velocity.

5. Spin $\hbar/2$ of the Dirac particle

The spin of the Dirac particle is based on the precession of its constituent vortices, mutually triggered by their rotational fields. The precession of a vortex *Y* within the rotational field of a vortex *Z* oriented orthogonally to it

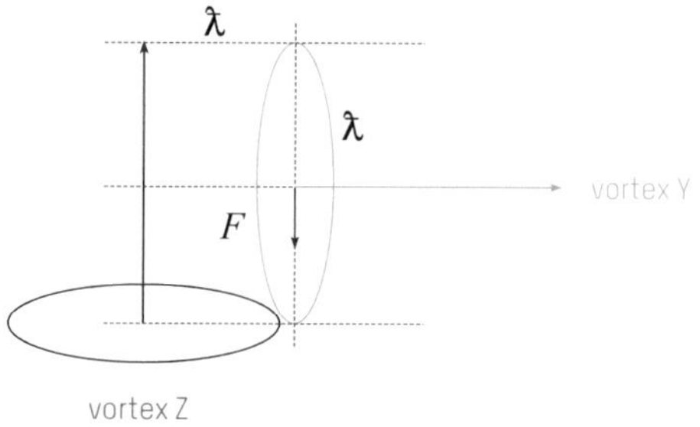

can be treated as the precession of an electron e^- within a magnetic field *B*, since the magnetic field is a rotational field. The frequency of the precession ω_p depends on the angular momentum *L* of the vortex *Y* and on the torque acting on it

$$\omega_p = \frac{1}{L}\frac{dL}{dt} = \frac{torque}{angular\ momentum}$$

The lever arm length for the *torque* is ƛ, and Maxwell yields the force *F*:

$$torque = \mathrm{F} \cdot ƛ$$
$$F = e\,(v \times B) = e \cdot c \cdot B$$
$$(torque = ecbƛ)$$

The angular momentum of the vortex is $\hbar = mƛc$, by which the precession frequency turns out as

$$\omega_p = \frac{1}{L}\frac{dL}{dt} = \frac{e \cdot c \cdot B \cdot \bar{\lambda}}{m \cdot \bar{\lambda} \cdot c} = \frac{eB}{m}$$

$$= \frac{e\hbar}{2m}\frac{2B}{\hbar} = \frac{2\mu_e B}{\hbar}$$

This corresponds to twice the de Broglie-Einstein frequency $\omega = E/\hbar$ with $E = \mu_e B$. Since the vortices have a frequency of $\omega_{vortex} = mc^2/\hbar$, by analogy, of the Dirac particle is $\omega_{Dirac\ particle} = 2mc^2/\hbar$, as computed backwards from quantum mechanical phenomena (p. 63). And since the radius can only be one half, $\bar{\lambda}_{vortex}/2$, due to $\omega_{vortex}\,\bar{\lambda}_{vortex} = c$, it follows

$$S = 2\omega_{vortex}\frac{\bar{\lambda}^2_{vortex}}{4}m = \frac{\hbar}{2}$$

for the spin of the Dirac particle.

It is thus the precession of its three vortices that determines the spin $\hbar/2$ of the Dirac particle.

ω_p itself is not caused by resonance, only ω_p is. It is the rotation of the field that continues B that is transmitted on the electron. This is expressed in $\omega_p = eB/m$ which contains no $\hbar$, or clearer still in the form $\omega_p \cdot m = e \cdot B$.

At the same time this explains why fermions are scattered as if their amplitude had an additional phase factor δ compared to bosons (for symmetry reasons the square $e^{-2i\delta}$ must be $=1$ and thus δ equal to π).

This realistic and local derivation of the half-integral spins of the Dirac particle proves (together with the three spatial axes which correspond to the three colours in the SM and with the spatial structure of the Dirac matrix) the tri-orthogonal vortex structure of the baryon, which lays the foundation for establishing all physics realistically and locally.

Appendix – Annotations

[1] The State of Inductive Physics.

Theory of relativity, electromagnetism and quantum mechanics are based on 13 mutually independent laws confirmed by experiment:

– momentum theorem, law of gravity,
– Lorentz invariance, equivalence principle,
– 4 Maxwell equations, law of electromagnetic force,
– 2 de Broglie-Einstein relations, uncertainty relation,
– Schrödinger equation;

as well as 7 mutually independent quantities, with values determined in experiment:

– 4 fundamental constants: c, G, h, α (fine structure constant),
– 3 masses: proton, neutron, electron (treated here as the "phenomena").

All further relations can be constructed from these 20 quantities and laws.

Cosmology still is open insofar as there

– is too little evidence for the constants in the Friedmann model (cosmological constant, mass density, radiation density, curvature $k = +1, 0$ or -1),
– are no explanations for the accelerated expansion of the universe,
– there are no explanations for the central acceleration within galaxies.

Elementary particle physics still is open insofar as

- essential theory has not been confirmed by experiment,
- an unsatisfactory large number of independent parameters is required for explaining few phenomena.

Elementary particle physics is represented by the Standard Model which says that

- a stable world is made from
  - electron,
  - *u*-quark,
  - *d*-quark,
- three quarks respectively form nucleons and are bound together by exchange of gluons:
  - two *u* and one *d* the proton,
  - two *d* and one *u* the neutron,
- all three elementary particles have spin $\hbar/2$ at rest,
- all three carry charge:
  - electron minus 1,
  - *d*-quark minus 1/3,
  - *u*-quark plus 2/3,
- charges act on one another by exchange of virtual photons.
- electrons do not participate in gluon exchange.

As *world in process* the Standard Model is more involved by one order of magnitude:

- muon and tau join the electron and form a family of charge minus 1,
- this family has an electrically neutral sidekick: the one of the three neutrinos,
- *u*-quark is joined by *c*- and *t*-, *d*-quark by *s*- and *b*-,
- additionally all quarks carry one of three so-called colours, and it takes three quarks of different colours for a nucleon,
- on top of these 4 times 6 particles come their anti-particles (inverse sign of all quantum mechanical parameters such as charge, spin),

- on transformation of one quark into another 3 additional exchange particles become active: one neutral as well as one positively and one negatively charged (together with the gluons carrying eight different colour transitions, plus the photon totalling twelve exchange particles),
- for assigning mass to the elementary particles Higgs field and its generating Higgs particle (detected in 2012 at the LHC in Geneva),
- for representing gravitation a thirteenth exchange particle is postulated: the graviton.

The Standard Model can represent all of the several hundred—from high energy laboratory, as well as in part from outer space—known particles as combinations of two or three quarks. Here next to the electron only the proton is stable. The isolated neutron decays, albeit relatively slowly. All others decay rapidly, some of them so quickly that they fail to complete one full rotation about their axis within their life span.

Summing up: With the Standard Model 51 additional independent quantities and laws relying on measurements come on top of the 20 laws and constants for "phenomenal physics" and the 4 of cosmology:

- 24 elementary particles (6 quarks times three colours, 6 leptons—of these already counting the electron under "phenomenal" physics; not counting corresponding anti particles),
- 13 exchange particles,
- Higgs field and particle,
- 4 coupling constants (3 of them sliding with relevant energy—at best converging; plus one for the confinement of the strong force),
- Dirac equation,
- 7 symmetries, Pauli exclusion.

In certain enhancements of the Standard Model with more exhaustive explanatory prowess the number of constants and laws can exceed 100 (super symmetry).

[2] **Epistemological Foundations**
in Hans Widmer: "The Concept of Uncompromising Humanism", rüffer & rub Sachbuchverlag GmbH, Zürich 2013
See extract p. 139ff.

[3] Quantum Entanglement

In so-called quantum entanglement, particle states correspond with each other across space and time without contact and without delay, which the laws of quantum mechanics (QM) predict, but contradicts perception ("spooky action at a distance", Einstein).

The contradiction is caused by the laws of QM not dealing with the causes of events, but with their statistics. Its predictions are like ancient peasants' weather proverbs, but like the peasant QM does not need to explain the causes in context. More clearly: As QM violates Bell's inequality[1964], any local reality is explicitly ruled out: QM is not a "realistic and local" theory.

Deductive physics is a "realistic and local" theory, since everything is connected with all via the continuum. At the same time it is a "statistic" one, since everything which materialises and can be measured is caused by resonance of interfering waves. Thus Einstein's expectations and the "Copenhagen interpretation" of QM are reconciled.

4 Concepts, Theorems of Physics

Concepts are a special case of information, and information is material structure—while all material reality is flowing. With DNA information enters into the world. For the first time one molecular structure is the instruction for the construction of another. The flowing is only recognisable qua information: it brings the flowing to a standstill like a photograph.

Before concepts and laws of physics can be exposed to testing they have to be invented as hypotheses. Kant on that in the prologue of the *Critique of Pure Reason:* "… that reason only accepts that which it pleads itself according to its design"; and Goethe adding with a tame Xenia: "What thou didst not put inside, that thou wilt never pull out."

The useful concepts for physics materialised late: mass for example only in the 17th, energy only in the middle of the 19th. Spin only at the beginning of the 20th century. Establishing these concepts was the first high invention.

Physics wants to grasp the fall of the apple from the tree—the process, not the location on the tree or on the ground. For this it arguably needs the still images, in its case literally: the states. It has to link two states by an operation: the altitude of the apple on the tree with the velocity on impact by the movement equation (mass m at height h in gravitational field g):

$$m\frac{d^2h}{dt^2} = -gm$$

Physics does not, however, content itself with that: integrating over dt it obtains

1. $$m\frac{dh}{dt} = -gm\Delta t + const$$

With $dh/dt = v$ (velocity) as well as $const = mv_o$ this turns into $mv - mg\Delta t = mv_o$. Thus it has won an invariant: the momentum.

If Newton's correlation is integrated over dh, it obtains a second invariant (with the initial height h_o, rearranged): the energy E involved in the process

$$2. \quad mgh - \frac{mv^2}{2} = mgh_o - \frac{mv_o^2}{2} = E_o$$

Momentum and energy are concepts capturing the process—but they are synthetic: neither materially observable nor envisionable—how should one envision for instance $mv^2/2$?

If instead of $h \rightarrow h + h_1$ is set in equation (1), the invariant does not change. With Hermann Weyl's (1885–1955) definition that if an operation on a quantity (original: on a "thing", e.g. turning a wheel) does not alter it, then the quantity is with respect to this operation "symmetric", momentum is symmetric with respect to translation in space. If instead of $t \rightarrow t + t_1$ is set in equation (2), this invariant does not change either: energy is symmetric with respect to time shift.

Emmy Noether (1882–1935) generalised these symmetry considerations in essence to the theorem: "a symmetry of interactions corresponds to a conservation theorem of physical observables." Thus symmetry became the most abstract law of nature—yet actually there is not more to it than what Newton put there. It is quasi the reversal of the invariant definitions. The eminently important vantage point for theory of relativity that $d(mv)/dt = F$ holds and not $mdv/dt = F$, does not follow from Noether.

There is no knowledge that no-one knows, no hidden laws of nature. All knowledge is confirmed hypothesis—e.g. by sixteen-year-old Einstein wondering what he would see when riding on a ray of light.

Theories never emerge from mere aggregations of data, but always start with something invented by the theory builder: for Newton with the ingenious concepts "mass" and "force", for Maxwell with his system of equations, for Einstein with Lorentz transformation and equivalence principle, for quantum mechanics with the Schrödinger equation. If according to Goedel's theorem rules of a theory cannot follow from itself, then they have to be put in by the theory builder. Einstein on this: "… the concepts surfacing in our thought … are all freely invented …"

The first great synthesis of physics was accomplished by Newton in 1687 based on Kepler and Galilei who had obtained their kinematic laws from observation. Newton's law did not follow from these laws, but he introduced "inertia", "gravitation" and "acceleration" newly, such that the former could be derived from them. Over this Einstein superimposed his laws in 1905. Einstein's field equations, too, can neither be read out of Newton nor the measured "constancy of the speed of light". His tensors are his invention proposed from outside, prompted by the Euler equation. Likewise Maxwell's equations, which he produced in 1873 on the basis of what was known about electricity and magnetism, come from the outside. Finally Schrödinger's equation established in 1926 is a new proposition.

Even if physics will one day be completed fully and free of contradictions (like electromagnetism since Maxwell), it will remain a construction of concepts and theorems whose only justification is that it stands the experiment. Concepts and theorems are approximations by the bounded ("defining") and the stationary ("understanding") to that which in reality is boundless and flowing.

Einstein

Newton — Michelson, Morley

Kepler — Galilei

Maxwell

Faraday, Oersted, Ampère — Licht

Elektrizität — Magnetismus

Schrödinger

de Broglie- Einstein — Newton

de Broglie — Planck

Dr Hans Widmer (1940) studied Mechanical Engineering at ETH Zurich and gained a PhD in Nuclear Engineering at MIT. After several CEO positions with international companies, he took over a company in the mechanical engineering industry as chairman, CEO and major shareholder. Widmer is married, father to four children and lives with his family near Zurich.

Extract

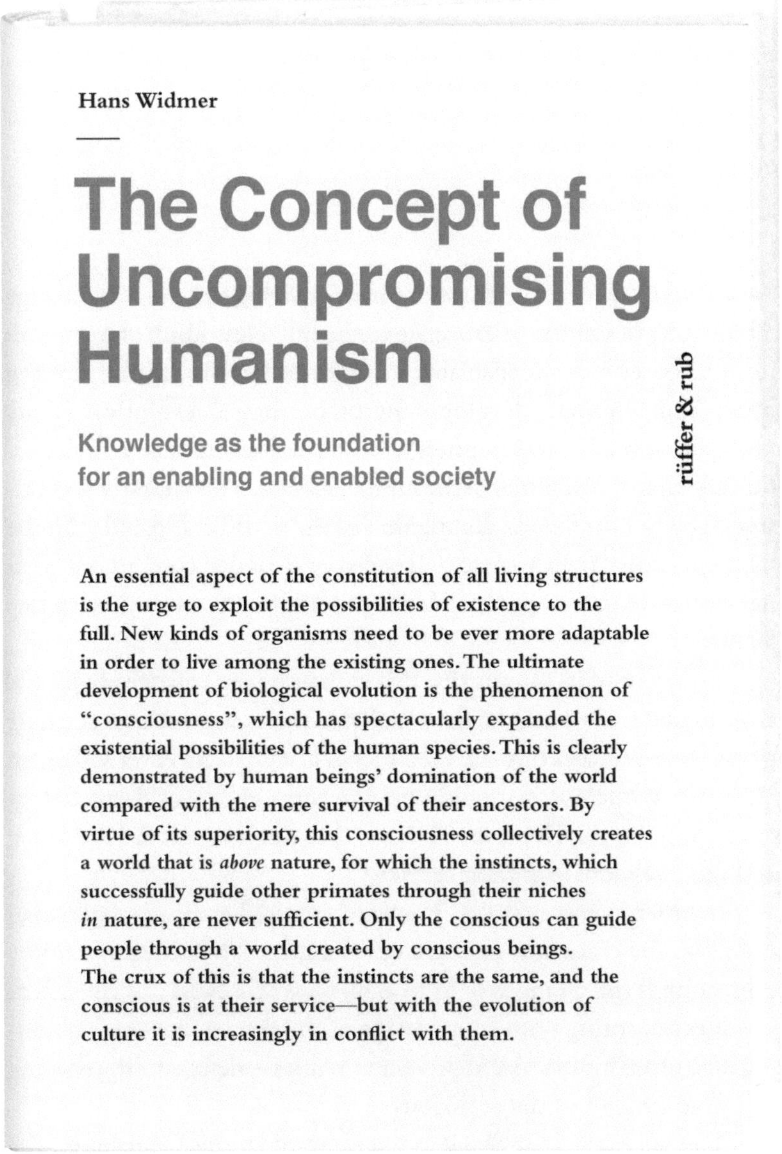

Hans Widmer

The Concept of Uncompromising Humanism

Knowledge as the foundation for an enabling and enabled society

rüffer & rub

An essential aspect of the constitution of all living structures is the urge to exploit the possibilities of existence to the full. New kinds of organisms need to be ever more adaptable in order to live among the existing ones. The ultimate development of biological evolution is the phenomenon of "consciousness", which has spectacularly expanded the existential possibilities of the human species. This is clearly demonstrated by human beings' domination of the world compared with the mere survival of their ancestors. By virtue of its superiority, this consciousness collectively creates a world that is *above* nature, for which the instincts, which successfully guide other primates through their niches *in* nature, are never sufficient. Only the conscious can guide people through a world created by conscious beings.
The crux of this is that the instincts are the same, and the conscious is at their service—but with the evolution of culture it is increasingly in conflict with them.

The cycle begins

An essential aspect of the constitution of all living structures is the urge to use the possibilities of existence to the full. New kinds of organisms need to be ever more adaptable in order to live among the existing ones. The ultimate development of biological evolution is the phenomenon of "consciousness", which has spectacularly expanded the existential possibilities of the human species. This is clearly demonstrated by human beings' domination of the world compared with the mere survival of their ancestors. It is also evident that consciousness has not been a tentative, gradual development, but an evolutionary *leap forward*.

By virtue of its superiority, this consciousness collectively creates a world that is *above* nature, for which the instincts, which successfully guide other primates through their niches *in* nature, are never sufficient. Only the conscious can guide people through a world created by conscious beings. The crux of this is that the instincts are the same, and the conscious is at their service.

In order to ensure that a "jungle of ever higher refinements" does not arise, the conscious must transpose instincts to a human culture. Humanity is still a long way from achieving this. Real human suffering does not spring from faulty design of the species, but—a frightening idea given historical and present atrocities—defects in the shaping and development of the conscious.

The necessary knowledge is put forward by this Concept on the basis of what science has produced so far, such as what is life, a human being, free will, happiness. Scientific knowledge is derived from the systematic questioning of what appears to be reality. Knowledge can only ever be obtained from such questioning. Philosophical work therefore starts with the incorporation of relevant knowledge: "The narrow gate that leads to wisdom."[*Kant*]

Knowledge compellingly leads to an enabling form of organisation of human societies—in other words to one that offers all its members the framework for a fulfilled life. It also invites individuals to use this framework to the full. However, acquiring this knowledge depends on the will to do so, and putting it into practice requires self-control, both of which emerge from it. An enabling organisation also requires those knowledgeable, self-disciplined individuals—enabled individuals—who are its product. This desirable state therefore cannot be decreed, but can develop in response to a catalyst—enlightenment.

Dissociation from conventional philosophy

Conventional philosophy starts from concepts such as "justice" or the "meaning of being". Kant even says derisively that it "gropes around for concepts". It does not care about reality, but about what philosophers have said. As the experiment is to science, so is the quote to philosophy. Philosophy seeks wisdom—a wisdom, however, that is not based on knowledge, but is a shot in the dark.

The hypotheses of philosophy cannot be proved scientifically—there are only philosophers, whose thoughts can be studied. They have published at least a million printed pages over the centuries, in which the majority—in line with the nature of such philosophising—are concerned with refuting others; some even turn their backs on their own published works. Thus even studying them all does not lead to universally applicable knowledge.

Structure of the concept

The Concept of Uncompromising Humanism makes no assumptions about anything in advance, nor does it presuppose any specific knowledge. It proceeds on the basis of perception and comes to conclusions using intuitive logic which readers can reconstruct for themselves, including concepts such as the theory of relativity or the hypercycle (the origin of life).

Before Kant, and to Kant himself, it was taken for granted that philosophy was able to acquire all available knowledge. However, at the start of the 19th century science began to move at a speed that philosophers could no longer follow—and even if they did, they were soon left behind by the theory of relativity and quantum mechanics. They fell back on "the clarification of propositions", [*Wittgenstein*] and some even resorted to mysticism, which is not what they should aspire to—rather, theyshould acquire the fundamental knowledge, move beyond it and perform their former tasks on the basis of this new, magnificent structure.

Consciousness radically expands the horizons of the conscious being—spatially, temporally and socially—especially as it includes the conscious being as the "self". As expanded horizons offer both possibilities and threats, the consciousness has to interpret *everything* it perceives within them. If it lacks knowledge, it manages with assumptions and assertions; but as world history makes clear, it is knowledge that brings the greatest success. The selection criterion in the obtaining of knowledge is freedom from contradiction, firstly with regard to reality and secondly with regard to all verified knowledge. Ensuring the maximum possible freedom from contradiction serves two purposes, the force of a statement *and* its completeness.

The inevitable conclusion is that a model explains the whole or it explains nothing. It is only complete if it explains not only the world, but also the thinking behind the explanations. Despite all attempts at didactic concentration, the quantity of knowledge to be processed for conclusiveness is extensive. This is not all—anyone who embarks on it will only know once they have made the investment whether it was worth it.

A closed circuit of statements? Sciences explain circumscribed reality in terms of circumscribed reality, and thus have by now built up an exponentially increasing, enormous body of knowledge; and yet no-one feels responsible for the consistency of the whole. Philosophers after Hegel have capitulated before the task, even sneering at those small-minded people who attempt it despite their warnings.

The basis for the Concept of Uncompromising Humanism is formed by the way in which the consciousness imagines the world to be; it ascends in stages towards a way of thinking produced by this imagining:

1. *A priori intuitions.*[Kant] Space and time ineluctably form the coordinate system in the human brain, within which it represents the world.
2. *Continuum, mass, cosmos.* A continuum logically and essentially fills the perceived space—from Anaximander to Einstein. In "deductive physics"*, on which the model is based, mass is derived as the dynamic of an appropriately-specified continuum, and this continuum carries the expansion of the universe.
3. *Atoms, elementary particles.* The interaction of the dynamics of masses phenomena and are the basis of all that is perceived. Elementary dynamics are structured into atoms, and atoms into inorganic molecules and, under appropriate circumstances, into organic ones.
4. *Life.* The huge accumulation of organic molecules on Earth led in a single event to a hypercycle of mutually-determining molecules: the basis for life. As far as is known to date, this has occurred only on Earth.
5. *Biological data processing.* In the course of evolution, the interaction between cells and cell structures was increasingly augmented by meetings between simple representatives of biochemical states—by means of signals in conductive pathways, ganglia, brains.
6. *Thinking.* Biological data processing gave rise to thought, which cannot avoid perceiving the world as a body in the coordinates of a priori intuitions.

The ontological circle answers the question "what can I know?", with which Kant's "Critique of Pure Reason" is concerned. However, beings with the benefit of consciousness do not have ontology as their primary aim, but happiness in their own lives. They demand answers to the kind of questions posed by Kant in his "Critique of Practical Reason": "How should I act? What can I hope for?", and to the question of how societies should be organised politically, economically, culturally. The basis for answering these questions is the certainty of the existence of *free will*. Kant postulated this without further ado, while current research into the brain questions it. The Concept of Uncompromising Humanism recognises it by its indispensable function, that of evaluating and selecting the solutions that thinking produces for the world, where instinct alone is not sufficient. In the space opened up by free will the possibility of successful life emerges—"happiness".

The Concept shows what successful individual life is and what leads to it, and also what are the conditions for it: the "enabling state" and its prerequisite, "enabled citizens". Each presupposes the other, both are derived from the preceding stages and form the foundations for the postulation that "all people could be equally happy". *Lichtenberg*

Ontological circle

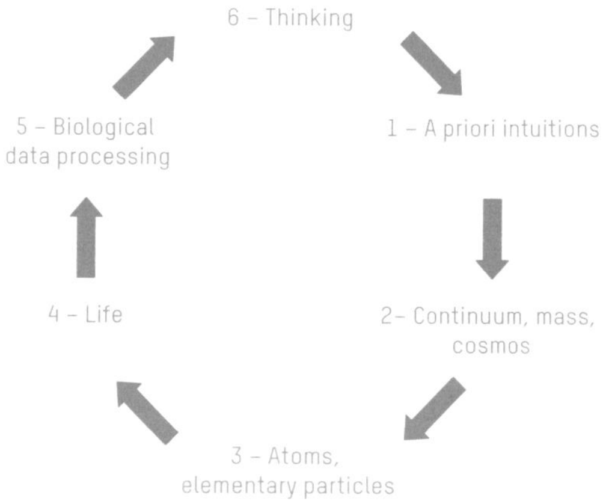

Each piece of scientific knowledge is based on a proposal that has "stood the test on the touchstone of reality".[Kant] In the Concept of Uncompromising Humanism, this proposal is the whole. It passes the test, as each of its stages are proven science and each stage emerges strictly from the previous one. It thus becomes clear that knowledge is that which is built up by consciousness and not, as Plato would have it, items in an inventory of knowledge that predated humans, to which they gain access gradually.

Hyperstases

The proposed Concept is confronted by two didactic challenges:

- A kind of uncertainty principle: the volume of knowledge needed for conclusiveness is incalculable; on the other hand, argumentation based on incomplete knowledge is not conclusive,
- The progression from one stage to the next.

Bypassing the uncertainty principle requires consolidation, visualisation and concepts such as *self-organisation, evolution* and *data processing,* which incorporate wide ranges of facts while at the same time retaining their essence. The quantity of individual items of knowledge does not prevent an overall picture from being gained, but is a prerequisite for it, as when completing a jigsaw puzzle.

In the difficult task of understanding the leaps from one stage to the next, it is worth recalling the following:

- A sandcastle is made from sand, but is not sand; it is a castle—something new that was not already present in the sand;
- A melody consists of notes, but its essence is not the notes;
- Life consists of molecules, but its essence is not the molecules.

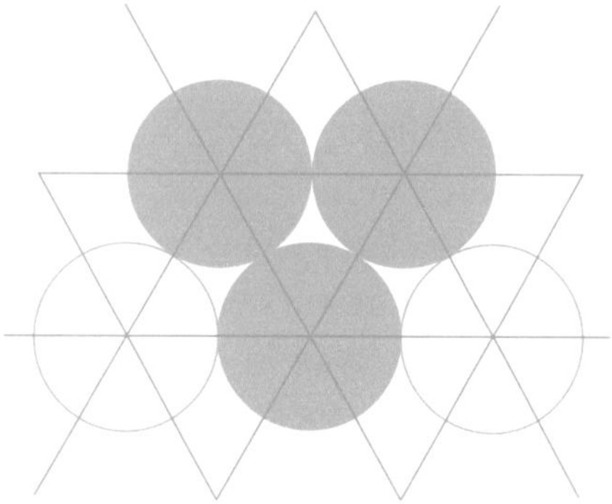

Add to this the phenomenon of *self*-organisation; if a load of gravel is tipped onto a building site, the debris forms a cone; this occurs of itself, as the cone was not thought out in advance. Similarly, if equal spheres are pushed against one another they organise themselves into equilateral triangles without any outside intervention. The Concept describes this self-organised evolution of the stages with the new term *hyperstasis*[I]: *hyperstasis = product of the self-organisation of a substrate.*

Inexplicable basis—six hyperstases

Human beings do not carry the world in their heads, but ideas of it, and the coordinate system for these representations are space and time. Anyone who, without any philosophical intention, asks what space and time are, will soon realise that it is impossible to refer them back to other concepts or to imagine they do not exist; they form the indispensable coordinate system for imagining the world. This insight renders unnecessary a range of philosophical questions, such as what is time or eternity, why anything exists at all and what is its purpose. Deductive physics removes the incompatibility between Einstein's theory of relativity and a priori intuitions.

Hyperstasis I: As a hurricane is formed from imbalances and consists of air and water yet is not air and water, but *dynamics* of them, so mass is the dynamic of the continuum. This is specified, while Anaximander's *apeiron,* Plotinus' One, Descartes' aether and Einstein's space-time continuum were only ideas. The mathematics needed to determine the behaviour of a continuum are field theories. All the major theories of inductive (conventional) physics are field theories, and they can be used to calculate, but not explain, the behaviour of everything from elementary particles through to galaxies.

Hyperstasis II: The combined effect of the dynamics of elementary masses leads again to something completely new: structures. This is because the rotation in space inherent in the dynamics of masses describes an axis (spin), an orientation which space, as a concept, lacks. At the lowest level in the hierarchy of stable structures are protons and neutrons, which in combination with similarly stable electrons form *atoms,* from which come *molecules,* and in favourable circumstances complex *organic molecules* (which are still not life). The science that describes the formation and cohesion of structures is called quantum mechanics. It arose of necessity from guesswork, however it evolves compellingly in deductive physics from the dynamics of masses, thus descending from the Olympus of the inconceivable in the same way as the theory of relativity.

Hyperstasis III: The essence of the leap to life lies in a cycle of structures, in which the positive of DNA is the blueprint for the negative and vice versa (hypercycle). In this way, the phenomenon of information enters the universe, on Earth according to the laws of nature, but apparently only a rare occurrence in the wider universe.

Hyperstasis IV: The combining of biological molecules into cells and of cells with each other is controlled by concentrations and differentiations: those which are to act on one another are in contact and those which are not are separated. The next great leap forward is to representatives of the forces emanating from molecules, to simple signals.

This is the leap to *biological data processing*—the ultimate source of all intellectual activity.

Hyperstasis V: The essence of the leap from biological data processing to *thinking* lies in decoupling certain aspects of data processing and making them independent from forces driven by reflex and instinct. This decoupled data processing constructs an image of the world, which in the infant soon becomes so all-embracing that it contains the subject itself. Once again we have a cycle: the subject thinks—thinking produces the subject.

Humans cannot be explained by the laws that govern life alone. What constitutes humans, i.e. thought or, in other words, consciousness, distinguishes them from other primates not just by degree, but categorically. With consciousness a phenomenon enters the universe that is as new as life itself. Consciousness is that expansion of the horizon which is the source of all joy and sorrow, all hope and fear, all that is human.

With consciousness comes free will as a concomitant, not as an additional hyperstasis. Humans are not free to choose into which world they are "thrown", and as what; rather, their freedom lies in the next step to be taken, and it is this freedom that they perceive. Similarly, "happiness" is a concomitant, of the physiological nature of learning: life-affirming intentions and experiences lead to the release of hormones that promote a positive mood.

Hyperstasis VI: The last of the hyperstases, culture, embraces innumerable human lives. Culture is more than the accumulation of individuals' behaviour—it produces language, society, the state, economy, science, art, philosophy and religion. These are all self-organised into historical periods, developing from barely-differentiated beginnings into distinct independent cultures, although the substrate always remains the same: human nature.

Genetically, humans have not developed further over the timeframe of human history. This was the starting point, for example, of Jacob

Burckhardt's description in "Reflections on History" of "humans, who remain unique, ... enduring, striving, acting, as they are, have always been and always will be". On the other hand, the religious, political and economic organisation of societies does evolve, defining the framework within which individuals, their consciousness and aspirations develop. Moreover, the 20th century saw substantial development of this framework with regard to human rights, democracy, education, health and welfare—despite all the century's retrograde barbarism. Nevertheless there is still a very, very long way to go before we achieve a culture worthy of human nature, that of "Uncompromising Humanism".

Humanism, to summarise, represents the striving for a way of life and social conditions that are worthy of our species. Humanism has been celebrated in thought and verse from Horace to the German Idealists of the 18th and 19th centuries, only to be most tragically defeated by reality: rather than high ideals, it was wars, genocide, communism, national socialism that prevailed. The celebration of the ideal subsided gradually, and more dramatically after the Second World War. The ideal of humanism was not a false one, but it is not enough merely to desire the desirable. "In matters of peace, talent and instinct play a more significant role than good intentions, which are of themselves totally characterless." *Robert Walser*

Uncompromising Humanism is the kind of idealism that begins with knowledge, the defining quality of humankind. Only that which has its basis in reality is viable—the idea of the world endorsed by the world. Proceeding from a priori intuitions to hyperstases, the Concept of Uncompromising Humanism inevitably results in the following: *individual happiness* need not fail any more than bold dreams, provided that human beings are enabled, know what there is to be known, and set their sights beyond the short term. *Societies are enabling* when individuals decide for themselves what is possible for them to decide; the same applies to the community, province, state; and states thus exist to serve the development of their citizens.

Since the Concept strictly follows reason, do its arguments not then lose sight of the "innermost being", the "divine", in each human being? Absolutely not, because reason

- provides a navigational instrument to help the innocent innermost being through the world created by human beings—the more viable the knowledge, the safer the path;
- teaches individuals not only how to find their way in the world but also to recognise and awaken the innermost being in its purity, wisdom and affirmation of life;
- in so doing reveals the divine in humans;
- shows the way, despite all the pressures and barriers that exist in the mind and in the world, towards the real, unfathomable, inalienable possession of a wise, stable, affirmative personality.

The difficulty here is that knowledge has to be acquired. If the love of knowledge in the world were as great as the love of God is in religious declaration, humanity would have progressed much further. In the words of Horace: *sapere aude* (dare to be wise).

1
Unshakeable foundations of all knowledge: a priori intuitions

Reality and representation

Ideas in the human brain are actively produced and are not mere reflections of the outside world. Visual images, for example, result from the processing of incident electro-magnetic radiation. Just how synthetic the picture is can be illustrated by an operation in which the optic nerves of a chameleon are switched: afterwards, it directs its tongue in precisely the opposite direction from its prey. The illusion is perfect: the subject considers itself an impartial witness of the presence of objects and demonstrates, as a matter of course, total reliance on the constructed picture. Goethe, on the other hand, in observing nature, constantly asked himself the question, "Is it the object or is it myself that is being expressed here?"

A photograph requires photographic paper, the molecules of which react specifically to wavelengths of incident electromagnetic radiation, for example to 400 nanometres for the reflection of violet light. However, the Mona Lisa can also be represented by the appropriate planting of grass in a field, or the course of a road can be drawn in the sand using the big toe. A substrate is required *for* a picture, and it is irrelevant what this is in itself; but *in* the picture there has to be order among the pixels. In a photograph, for example, there are no spatial relationships, but only two-dimensional ones, which the eye reconstructs into objects in space based on the laws of perspective. A drawing in the sand ("here is Rome and here is Paris") implies scale and a north-south axis; the sequence of images in a film requires the addition of a temporal order.

This is true not only for visual images, but for all ideas—however the human brain constructs its pictures, they have to offer reliable help in our perception of the world. Thomas Aquinas sums it up perfectly: "An object in the mind is adopted according to the mind—and not according to the object."

If it is understood that

- space can only be experienced through movement—and therefore in time; time similarly only through movement—and therefore in space;
- only bodies can have such "experiences",
- bodies are characterised by the fact that they are permanent (in time) and impenetrable (in space),

this is no circular argument in which the hypothesis already contains that which is to be proved, but rather an expression of the nature of the process of representation which is only concerned with the correspondence of relations. Bodies, space and time are not reality, but the phylogenetically-provided means of producing an idea of reality.

The fundamental and literal impossibility of grasping time and space led Kant in 1781 to introduce the concept of "a priori intuitions" to philosophy: "Space is not experience, for all spatial experience assumes the idea of space." And "Time is nothing other than the subjective condition under which all intuitions can take place within us."

Kant had trouble with "body", or "substance" as he called it. He did put forward a "principle of the permanence of substance": "All appearances are in time ... In [them] the substrate must be found ... [which] is permanence ... Therefore in all appearances, permanence is the object itself ..." He did not, however, establish the link between substance and space (impenetrability) that is analogous to this substance-time relationship (permanence). He lacked the concept of atomism which sees all matter as made up of the smallest units, and was therefore unable to classify substance in its "multiplicity of appearance" as an a priori intuition, although he wrote about it as if it were possible.

It is not surprising that Kant had difficulty with the concept of substance, for physics is equally unable to define what substance in the sense of "mass" is. One knows how "heavy" a mass is, in the material sense, from the pull the Earth exerts on it—mass multiplied by gravity—and how "inert" it is by its resistance to acceleration—mass multiplied by acceleration. When equating the two, mass is removed so that acceleration is equal to gravity, which is the same for all objects of any mass, from goose feathers to lead bullets. This only establishes the gravitational field of the Earth, that is the interaction of masses—not what mass actually *is*.

Physics simply treats mass as a given and uses its behaviour to derive force fields, which are mathematically expressed in field theories. All field theories contain a continuum, in Einstein the "space-time continuum". Thus, conventional physics proceeds from this inductively, namely by inferring a continuum from the phenomenon.

Deductive physics takes the opposite route, from the continuum to mass as the dynamic of it. The constituents of this continuum are pure bodies, defined as permanent, impenetrable volumes, the opposite of empty space and characterised by nothing else, which is why deductive physics adds "body" as a third a priori intuition in addition to those of "space" and "time". Hence, all that is needed for the representation of the material world is the system of coordinates covering time and space, and the bodies within it, all material phenomena being derived from them.

The three a priori intuitions have a correspondence in the three fundamental physical constants, which would confirm philosophy as well as science in their claim to be the basis for all knowledge—if a century ago the theory of relativity had not come up with ideas that declared a priori intuitions invalid: with time being stretched, space being stretched and curved, and with mass increasing as its own speed increases, tending to infinity at the speed of light.

The triumphant proof by experimentation of Einstein's predictions undermined philosophy, and in order to regain a firm footing it was necessary to get to grips with the theory of relativity. As children begin their development without consciousness of space and time as dimensions independent of their own existence, the first thing to be observed is how this emerges as a capacity for abstraction and objectification.

What are the benefits to a person of knowing how matter can be thought about? Firstly, the natural reflex of wanting to have everything explained is satisfied. It can be built on, and an explanation for "life" found in a few steps. It is now possible to understand "intellect", the essence of which, although it may go beyond matter, is nevertheless a structure of matter. Finally there comes freedom from all speculation, freedom to construct one's idea of the world on the solid foundation of recognised laws.

Deductive Physics

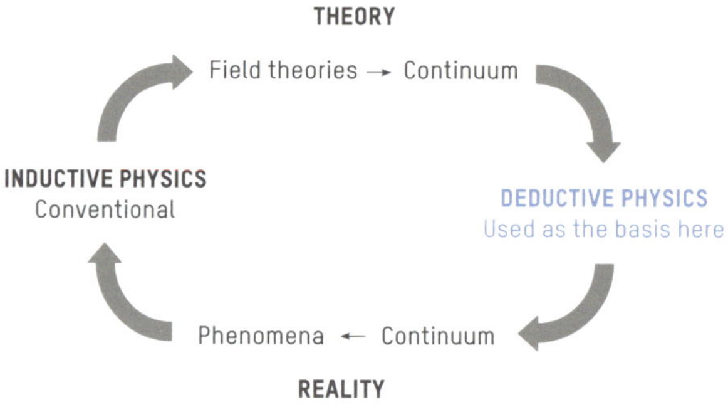

Acquiring space-time concepts

With freedom of movement comes the need to explore the space where it occurs. Here it is only a matter of distinguishing between "space" and "impenetrability", not of what the impenetrable barrier *is.*

Infants' first experiences of touch establish impenetrability: is there a body in the way of the hands or not? Playing with wooden bricks teaches them that two objects cannot occupy the same space at once, and conversely one cannot be in two places at the same time. By the age of eight months they have internalised the concept of permanence; if an object is covered they look for it, whereas previously they would simply have turned their attention to something else.

At first, space manifests itself only as the distance from the child to an object; later the perspective widens with the awareness of differences in length. Time is initially understood in terms of earlier and later, faster and slower, longer and shorter. If one of two trains running on parallel tracks in a model railway is faster than the other, a toddler will see it as going further, without the ability to express the idea that it will arrive at its destination sooner.

Children cannot conceive of objective time and objective space, both existing independently of the child's own presence, until the age of seven or eight—and from then on they can never again imagine them not existing. They are inescapable, and yet philosophy was stalled in 1919, when an experiment during the solar eclipse in England confirmed Einstein's mathematics, which he construed as a sequence of the stretching and bending of space and time, something which from the perspective of deductive physics is unnecessary.

Fundamental constants

Inductive physics represents the material world in space and time, but instead of the "body" dimension it relies on "mass". It defines mass in terms of a specific *volume* of a specific substance: a litre of water is a kilogram, and all substances that are similarly inert and heavy are the same. All statements made in physics are stated using the three dimensions of length (for space) in metres, *m*, time in seconds, *s*, and mass in kilograms, *kg*. Electricity is linked to mass by means of the dimensionless fine-structure constant α, and does not represent an additional dimension. Also, as physics uses three dimensions to express itself, there are three fundamental constants[1]:

c, the speed of light $\frac{m}{s}$

G, the gravitational constant $\frac{m^3}{kg \cdot s^2}$

$\hbar$, Planck's constant $\frac{m^2 \cdot kg}{s}$

Metres, seconds and kilograms are arbitrary measures: a 40,000,000th part of the equatorial circumference, an 86,400th part of a day, the inertia and weight of a litre of water—while in contrast the fundamental constants *c, G, ħ* are facts, and are what they are independently of the measurement units of physics. If they had different values, the world would look different: elementary masses would be larger or smaller, or there would be none at all. Gravitation would be so strong that all celestial objects would be drawn together into a single lump, or would be so weak that nothing would hold together. Quantum mechanical interferences would be so weak that electrons would fall into their atomic nuclei, so that no molecules and no life could arise, etc.

As in deductive physics the fundamental constants are the properties that determine the continuum in space and time, and as all materials are derived from the dynamics of this continuum, all phenomena are based on a priori intuitions and therefore assumed "according to the mind". *Thomas Aquinas*

Space and time coordinates run to infinity, revealing their nature as concepts. On the other hand, the universe as portrayed within these coordinates is seen to be finite. In retrospect the two giants can be reconciled: Newton's "absolute space" and "absolute time" relate to the coordinate system of all representation, while Einstein's "absolute speed of light" relates to the continuum represented within it.

Irritation from the theory of relativity

Kant's a priori intuitions are the most fundamental examinations of thought that philosophy has produced, but at the same time they are the most persistently refuted: thinkers are constantly putting forward new speculations, in particular about the nature of time.

In 1905, Einstein introduced the terms "expansion of time", "time dilation" and "space-time continuum", adding "curved space" ten years later, thus causing bewilderment and relief in equal measure: bewilderment for those who believed they understood the concept of a priori intuitions; relief for others as there was now something much more inconceivable, so it must be the truth.

The special theory of relativity goes beyond any intuition right from the start of the derivation: a "four-vector" is first introduced for spaces, then rotated around an imaginary angle, and later it is concluded formulaically that impulse is also a four-vector (with time in the fourth dimension)—and after a chain of abstract operations, $E = mc^2$ is obtained. A professor* at the Swiss Federal Institute of Technology (ETH) once said to his students, "You go through it step by step, accept what emerges, and understand nothing. No-one understands it."

In his lectures at Princeton in May 1921, Einstein made fun of the fact that physicists had been obliged to "bring down ... the concepts of time and space ... from the Olympus of the a priori ..." He was apparently confusing "a priori" with "absolute" and failed to appreciate that Kant's a priori intuitions identified a more radical relativity than his theory of relativity, i.e. that between thought and reality—not merely that between two bodies moving relative to one another (special theory of relativity) or interacting with one another (general theory of relativity).[2]

Knowledge grows from the resolution of contradictions, and the contradiction first resolved by Einstein was this: if a source of light moves towards an observer at speed v, and the light is moving away from the source at the speed of light c, the observer intuitively expects an arrival speed of $c + v$. But in the 1880s, it was determined by measurement that the velocity was c in all circumstances (Michelson and Morley). How did Einstein resolve this contradiction? His very first step contains all the irritation of the later results: he said to himself that if the speed of light is to remain constant at c regardless of the position of the observer, then distance and time simply have to be "relativised". Instead of Newton's space and time, he therefore proposed c as the absolute. He then tested out how a system of coor-

* Jakob Ackeret, 1898–1981; physicist, pioneer of fluid mechanics; proposed the term "Mach number".

dinates *K'* with its origin in the light source would have to relate to the observer's system of coordinates *K*, to enable light both to be emitted from there at *c* and to be received here at *c*.

His conclusion was that space and time are contracted around the light source, but the consequences go much further: mass increases as *v* increases and thus also the momentum (momentum = mass multiplied by velocity). A momentum[3] has an energy and a cross-multiplication supplies directly the result of the century—that this energy is not zero, even at rest, but the famous $E_{rest} = mc^2$.[4]

Einstein would have been severely taken aback had he realised that his result originated in Newton's formulation of the momentum conservation law: if Newton had simply written "force equals mass multiplied by acceleration", Einstein would not have made such a leap forward. He was merely lucky, since in 1905 there was as yet no experimental proof that Newton's intuitive formulation applies.[5] But even leaving this aside there was great cause for amazement as kinetic energy was now to be understood as a pure increase of something that no-one had bargained with: rest energy mc^2. It is an indication that that mass is a dynamic, not a corpuscle.

The simple reason for this, that light from any source is radiated at *c* and received by any mass at *c*, regardless of whether they are moving relative to one another, depends from the point of view of deductive physics on the fact that

- the continuum directly on the surface of a mass is at rest (in the same way that air is at rest on the outer ear despite the strongest wind—it does not blow into or through the ear),
- the speed of propagation of all disturbances (such as waves) in a continuum at rest is *c*.

However the frequency of lightwaves hitting the observer should not merely be expected in terms of linear addition (the original quantity of signals per second plus the gain from the approach[6]), as the field of a mass, spherical when at rest, is contracted if it moves at *v* relative to the continuum—like a source in the countercurrent. This shortens the wavelength of the radiation by the factor known as the Lorentz contraction, and the frequency is increased in inverse proportion, which leads to the Doppler effect[7], in which the frequency increases at more than a linear rate, and for $v \rightarrow c$ becomes infinite (approximately corresponding to a sonic boom). This enables all the results of the special theory of relativity to be understood and also some of those of the general theory of relativity, if kinetic potential is replaced by gravitational potential in the formulae. The assumption of a continuum and the representation of a mass dynamic within is therefore sufficient to avoid the concept, incompatible with thinking, that space and time can expand and bend.

The predictions of the theory of relativity are true, but Einstein's interpretations of the correct mathematical results as the expansion of space and time should be replaced:

- It is not that the time of the mass in motion moves more slowly, but that its signals take longer to reach the observer,
- Space does not expand or contract, but the continuum in space, similar to the air that flows around a body,
- It is not the mass that increases with velocity, but its effect—similar to the pattering of rain at high speed against a windscreen,
- Mass should be thought of as a dynamic, as compelled by $E = mc^2$, and the idea of literally inconceivable corpuscles should be abandoned.

Ultimately, the theory of relativity only formalises the relativity of interactions: as a motorbike approaches a listener, the listener registers higher-frequency sounds, and as it goes away, lower. The theory of relativity does not offer anything else to philosophy, although it introduced a new era in physics.

Lorentz contraction

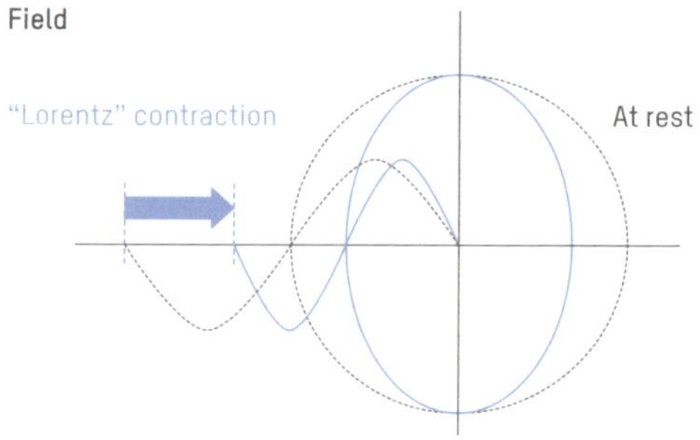

Irritation from quantum mechanics

As an infant's consciousness first begins to develop, there is an undirected movement of the limbs until an effect is achieved; after several repetitions this is then stored as an action-effect pattern. The action does not result from a physiological need, but a reflex that trains the brain. The pattern contains the idea prior to the triggering of the action, in the same way as a bird "imagines" the landing before settling onto a branch.

In this way, infants register their own intentions, and at nine months children recognise their intentions to such an extent that they can also recognise those of other people. Others are understood per se as intentional beings with intentions analogous to those of the child. This is manifested in the form of pointing to things and persons, in other words in attracting people's attention, something that is not observed even among the most attentive of the other primates. *Tomasello* By analogy, children later attribute intention to all processes, saying things like, "The ball wants to come to me" and seeking intention everywhere: "Why does a cherry tree want to blossom?" Theories of cause are also stated: "The moon shines so we can find our way home." The history of ideas began in similar fashion: mythologies invented beings with intentions in response to questions of cause and purpose; religions responded with creation stories.

A child of western civilisation gradually learns to transpose intentionality into causality and explain reality from reality. This was the giant step forward made by the pre-Socratic philosophers with the causality principle: "Everything has a cause" and the law of cause and effect: "Equal causes have equal effects".

One of the first experiences of causality a child encounters is that a body which was there first must be removed if another is to take its place. Kant considered causality to be a priori; however, it is not to the extent that, in the final regression, it is attributable to the fact that space can only be occupied by *one single* body, and so is already contained in the three a priori intuitions of space, time and body. Causality describes sequences of conditions, of stationary images, the earlier of which are termed causes and the later, effects. The stationary images are subjective constructs—considered objectively, "everything is in a state of flux", one thing flows from another, and in this sense everything that happens is from the outset "causal".

At the start of the 20th century physical experiments were advancing into atomic dimensions, revealing an acausal, inexplicable world. In the 1920s a handful of brilliant physicists developed quantum mechanics, with which all the probabilities and inexplicable conditions could be calculated—but not explained, for which reason the uncertainties are stated to be objective and the probability calculations raised to the ranks of fundamental laws of nature. The expectation of causality at the root of all phenomena was challenged. Philosophy was dumbfounded, and conventional physics was expanded by a further colossal dimension in addition to the theory of relativity.

In deductive physics, all quantum-mechanical facts arise from interferences of the waves that are emitted by masses and transmitted by the continuum. The specific values are shown to be resonances—like the vibrations in musical instruments—and the uncertainties consequences of the fact that interactions occur in waves and it cannot be determined precisely where the masses from which they originated were located in the wave. Deductive physics thus explains all phenomena causally, yet the information on an atomic scale is never sufficient for more than the calculation of probabilities, for which quantum mechanics provides the perfect instruments.

Philosophers have used the quantum-mechanical facts to perform "bold works of genius",[*Kant's expression*] going as far as an explanation of free will, although quantum phenomena are no more significant than others that can only be determined statistically, such as the behaviour of gas molecules (thermodynamics) or traffic. In everyday life there is much that appears to be acausal, or "chance"—we meet a neighbour in a remote place, or lightning strikes. The chance element here is that we had not taken account of our neighbour's trip round the world or of the electrical discharges from the sky. Both had causes, but neither had any kind of intention, concepts that are easily confused in everyday life. The thermal movements of the individual molecules of a gas also have causes, but they cannot be computed mathematically. But there are populations of them, which lead to the thermodynamic laws with the mean values of density and temperature.

Causes are frequently sought at too high a level, for example, there is no "cause of traffic", only causes for the individual drivers or passengers. Similarly, there is no "cause of human beings", merely a practically infinite number of evolutionary stages.

The categorisation as "chance" is therefore always based on a lack of knowledge and represents that which someone had not taken into account, the incalculable (Einstein: "The point where our calculations fail we call chance"), or that which has come about as a hyperstasis from a substrate underlying the phenomenon.

Causality is not equal to determination ("anonymous intention"). The latter would require a plan that "knows" the future outcome in advance and seeks to achieve it—from the smallest to the universal scale. But such plans are inconceivable, because they would contradict fundamental knowledge, in particular the development of thought from biological data processing and the mind from thought.

Could there not be a power in the universe, hidden from man, which acts, connects, steers, causes someone in London to cry out in the middle of the night only to discover subsequently that at that moment his brother in Alaska had died? There is much that is unexplained, but nothing is to be gained by explaining this away with the inexplicable (such as "there must be something out there"). It remains inexplicable until it is explained. The amount that is hidden is limited only by the imagination. But as long as it remains hidden, it cannot be used as an explanation. At any given time, the structure of knowledge only contains what has been discovered to date, and to talk of anything beyond that is in vain.

Long-distance effect, continuum

When children are asked what the sun is made of, they say something like: "From tiny shining pieces of cloud"; or where sugar dissolved in water has gone: "It's in pieces so small that you can't see them." The arguments of the pre-Socratic philosophers were similar: they got as far as indivisible particles from which everything is made. Anaximander called this *apeiron,* while Heraclitus added that it was constantly in a state of flux and Democritus went further by saying that everything real was made up of compounds of it. With hindsight it can be seen that the continuum was no quirk of Anaximander, Plotinus, Descartes or Einstein, but an expression of the nature of thought.

Einstein clearly assumed that Descartes' aether was immutably linked to Newton's absolute space, which is not compatible with the "constancy of the speed of light", and this is why he rejected the idea of an aether. If it is allowed that the aether (or whatever name is given to the continuum) can also flow, there is no longer any incompatibility with the a priori intuitions. At the latest when it comes to the expansion of the universe it is no longer possible to avoid an acceptance of flux.

It is logically essential to think of space as filled with a continuum when seeking to examine long-distance effects: if A is to have an effect on B, A must be in contact with B. Infants demonstrate that they understand this concept by pulling on a sheet in order to be able to reach an object that is resting on it. Descartes ascribed the transfer of effect to his aether; the constituents of the aether would collide, thus passing on momentum. All field theories state simply that: the field connects causes and effects through contact at extremely small (in mathematical terms: "infinitesimal") intervals.

Since Anaximander the continuum has been an analogy for air, a gas consisting of molecules with potential—manifest in its continuous motion—and space. The continuum is and remains the "One" which cannot be further divided, like the sand in a sandbox from which children build castles.

Not only is it an essential prerequisite of empty space that there are permanent, impenetrable bodies—or the concept of space would be pointless—, but these can and must move, or the concept of time would be pointless. The continuum fills the framework provided by a priori intuitions.

If the continuum were a gas, the fundamental constants c, G and $\hbar$ would represent temperature, reciprocal density and free path length. To reiterate, the constants as such do not have mass, but only volume, distance from one another and movement. Mass is only constituted through their dynamics.

What is the meaning of "existence"?

If all phenomena are traced back to bodies, their "existence" is determined by the existence attributed to the a priori intuition of "body". Permanence and impenetrability say it all, and "existence" turns out to be anticipated by the a priori intuitions.

What is the existence of the mind? In terms of "matter", the mind is information, the essence of which does not lie in the corporeal, but in structures. If it is understood that by the impenetrability of bodies it is not any fact that is meant, but rather an interaction, in other words the extent to which bodies block the way of other bodies, push them around or are pushed around by them—in the same way as small children discover bodies—, then by analogy it becomes clear that the "existence" of information also means the effect of information. Information is the structure that is communicated.

Physics talks of electrical charge, although it cannot say what constitutes charge, but rather measures interactions; for these it assumes charge to be the underlying cause. Charge is a property of elementary particles and has no isolatable existence, just as acid is its acidic effect, the wall perceived by a bat is something it cannot penetrate, and a "house" is an expression of its sheltering function. The existence of human beings is what they cause and tolerate—not their biomass.

In abstract terms, *"existence"* means *"to be in a state of interaction"*. "Existence" grows from linguistic usage: it is said of an objectivised object, thus one that is separate from oneself, that it "exists". The object is part of the inventory of the speaker's world—what is actually being expressed is that the speaker has noticed the object. Existence, or the verb "to be", can thus be replaced in the following sentences: berries are/glow red; schoolchildren are/sit in the classroom; two and two are/make four. The fact that something is or exists as a projection of the speaker is, like all verbal expressions, only justified by its usefulness.

Limits of knowledge

Those who want to conceive of reality must also consider their capacity for knowledge as part of objectifiable reality, beginning with the understanding that continuum, space and time are not part of the picture, are not reality, but the material and framework for forming it—sand and sandbox with which a model of the world can be built up.

The a priori intuitions should be accepted and not interpreted further. So why all the fuss about them? Because the framework they mark out for all philosophising cannot be transcended, even if the intention is to consider "outside" aspects. Philosophy was clearly too uncertain of the a priori intuitions to call Einstein back to the drawing board when he postulated the bending and expansion of space and the expansion of time. "Nature has given us the chessboard beyond which we cannot operate …" *Goethe*

Copernican turning points

Ask a three-year-old who has a brother whether the latter also has a brother and he will answer no, there are only two of them. Not until a year later will he come to consider his relationship with his brother from the outside, thereby objectivising it. Small children also relate the rest of the world to themselves, for example, when they say the moon shines so they can find their way home. The cognitive development of the human race, as well as individuals, is characterised by increasingly letting go of a viewpoint defined by how the outside world affects us.

Animism in hunter-gatherer cultures projects intentions onto everything that happens in the world—clouds as an expression of the moods of the gods, these in turn being a response to human behaviour. The *first turning point* in intellectual history was initiated by the pre-Socratic philosophers (600–400 BC): they considered world events as something apart from divine intentions, and justified reality with reality, although spatially as well as spiritually people were still at the heart of things. This world view, to which Ptolemy (100–170 AD) put the final touches, held its ground and was supported by scholars for some 2000 years.

It was Copernicus who introduced the *second turning point:* the universe did not revolve around the Earth, but the Earth around the sun. His grounds for believing this were because it made it easier to understand the movement of celestial objects: "Any visible movement of the firmament with its fixed stars is not due to movement of the firmament, but as it is seen from the Earth ..." The Earth and the human race were thus shifted from the centre of all cause and purpose—a colossal attack on the word of God, the authority of the church and the understanding at that time of the self and the world. To begin with, Copernicus was not taken seriously by very many (Luther considered him a fool), but then, sixty years after the publication of his work, Giordano Bruno was burnt at the stake for making the same statements.

In 1905, Einstein's theory of relativity introduced a *third turning point,* although this upset the a priori intuitions. It does not show that physics is thought out within the limits of human cognitive capability; nor does it show that it is a collection of abstract concepts and principles to explain the aspects phenomena have in common.

It has always been apparent contradictions that have led to new findings, in particular to a higher plane of objectivisation:

- Copernicus asked: "Why do the planets not move in the same way as the other celestial objects?",
- Einstein asked: "Why does light arrive at c, when it would be expected to have been emitted at $c + v$?" (where c is the speed of light, and v is the approach velocity of the emitting light source),
- In deductive physics the equivalent would be, "If the a priori intuitions are an immutable part of the thought process, how can the results of the theory of relativity be brought into harmony with one another?"

Notes for "The cycle begins"

1 Plotinus named the stages from which everything emanates "hypostases", but without considering the aspect of self-organisation or emergence from one another. The word is made up of "hypo", a basis, support, substrate, and "histánai", to set, place, lay. The similar term "emanation" does not include a state of self-organisation, and "autopoiesis" relates to narrowly-defined systems. This is why the new term, "hyperstasis", has been coined.

Notes for "1—A priori institutions"

1 Fundamental constants. *c*, *G* and $\hbar$ are represented in the three dimensions of length (for space) in metres *m*, time in seconds *s* and mass in kilograms *kg*:

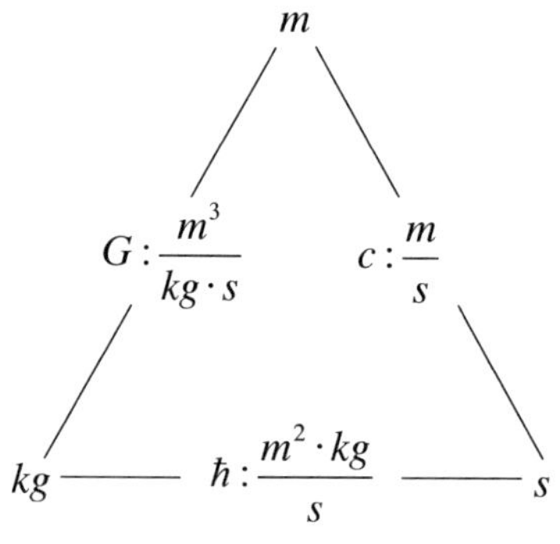

Charge *e* from the fine-structure constants

$$\alpha = e^2/\hbar c \rightarrow e = \sqrt{\alpha \hbar c} = \sqrt{\hbar c/137}$$

Conversely, the three dimensions can be expressed by the fundamental constants:

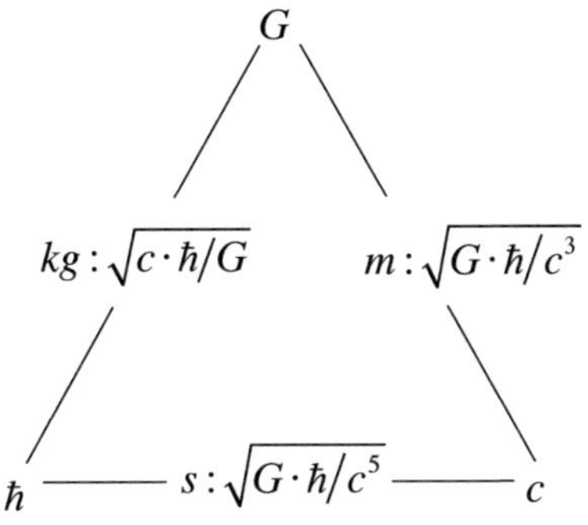

These radical expressions are given names such as "Planck length", and are understood in string theory as vanishing points—"the smallest meaningful length", as much as a factor of 10^{19} smaller than the diameter of a proton. Conversely, the "Planck mass" is heavier than a proton by the same factor, 10^{19}. It represents the theoretical mass that equates to a black hole:

Compton length $\frac{\hbar}{mc} = \frac{2Gm}{c^2}$ Schwarzschild radius, which gives

$$m = \sqrt{\frac{\hbar c}{2G}}$$

2 *Derivation of the results of the special theory of relativity*—without their formal apparatus; it is more instructive with Pythagoras and cross-multiplication:

Dilation of space and time. The simplest approach assumes that a point at rest in K moves to K' at $-v$, and $x' = x - vt$ is applied in accordance with Galileo. Then (because of the condition that where $x/t = c$, it must also be $x'/t' = c$), $t' = t - vx/c^2$. The calculation does not work, however, for eccentric relative movement (past the origin of K at interval r); but if an attempt is made with a factor a, $x' = a(x - vt)$ and $t' = a(t - vx/c^2)$, a in accordance with Pythagoras is found to be $1/\sqrt{1 - v^2/c^2}$ and thus

$$x' = \frac{x - vt}{\sqrt{1 - v^2/c^2}} \qquad t' = \frac{t - vx/c^2}{\sqrt{1 - v^2/c^2}}$$

from which the essential aspect is immediately seen: the light source can be ascribed a "proper time", i.e. the t' at its origin, where $x' = 0 \rightarrow x = vt$. Substitution gives $t'_{proper} = t\sqrt{1 - v^2/c^2}$—the inevitable consequence of the idea that the coordinates must give way if c does not. Einstein clearly showed the lower value of t'_{proper} to be a lower frequency, when he stated in Princeton: "A clock at rest at the starting point, the ticking of which is characterised by $l = n$ (sic!) therefore, viewed from K', goes slower at speed $l' = n/\sqrt{1 - v^2}$ ..." ("real light time", $l = ct$; "v" $= v/c$).

From the start, the variable t represents duration and not frequency, relates to intervals, and $t' < t$ therefore expresses that time intervals appear shorter. In fact it is the case that the (optically-transmitted) ticking at frequency ω of a clock in K reaches K' at a higher frequency: i.e. at $\omega' = \dfrac{\omega + kv}{\sqrt{1 - v^2/c^2}}$.

If $\omega/k = c$ is used for light, the product of the shortened interval and higher frequency is $\omega'\Delta t' = \omega\Delta t(1 + v/c)$—and the flow of time takes the intuitively-expected course.

Often muon decay is quoted to confirm Einstein's assertion, since its life increases at $1/\sqrt{1-v^2/c^2}$—but it increases with the energy (and thus the frequency) that change with the reciprocal radical expression.

3 *Relativistic momentum.* The special theory of relativity asks how one moving with an object should estimate their own momentum; this refers to the momentum relating to *K*—as in relation to *K'* it is at rest. But the time should be measured with their clock, *p'*:

$$p' = m\frac{dx}{dt'} = m\frac{dx}{dt}\frac{dt}{dt'} = mv\frac{dt}{dt'}$$

Using the previously-defined proper time t' gives $p' = \frac{mv}{\sqrt{1-v^2/c^2}}$.

4 $E = mc^2$. Instead of following the chain of abstract operations of the special theory of relativity which led to the result of the century, let cross-multiplication be used to determine the energy *E'* that belongs to this *p'*: for Newton dp'/dt is a force, as is dE'/dx. Thus, $dx/dt = v \; : dE' = dp' \cdot v$. The derivation of *p'* according to *v* results in $dp' = mdv/\sqrt{1-v^2/c^2}$.
Substitution and integration gives $E' = \frac{mc^2}{\sqrt{1-v^2/c^2}}$.

5 If, instead of $d(mv)/dt = Kraft$, Einstein had used $mdv/dt = Kraft$, still considered equal in 1905, this would have given $dE = mvdv$ and therefore for the energy only $E = mv^2/2$—not $E = mc^2$.

6 Linear frequency superposition $\omega_{observer} = \omega_{emission}\left(1 + v/c\right)$.

7 Doppler effect $\omega_{observer} = \omega_{emission}\frac{\left(1 + v/c\right)}{\sqrt{1 - v^2/c^2}}$.

For example, if v is half the speed of light, the frequency perceived by the observer is not simply 1.5 times that of the sender, but 1.73 times. Indeed, if $v = c$, the frequency would tend towards the infinite.

Content